WITHDRAWN

ALGEBRA FOR ATHLETES

ALGEBRA FOR ATHLETES

CAMERON BAUER

Nova Science Publishers, Inc.
Huntington, New York

Editorial Production:	Susan Boriotti
Office Manager:	Annette Hellinger
Graphics:	Frank Grucci and Jennifer Lucas
Information Editor:	Tatiana Shohov
Book Production:	Patrick Davin, Cathy DeGregory, Donna Dennis, Jennifer Kuenzig, Christine Mathosian, Tammy Sauter and Lynette Van Helden
Circulation:	John Bakewell, Lisa DeGangi and Michael Pazy Mino

Library of Congress Cataloging-in-Publication Data

Bauer, Cameron

    Algebra for Athletes / by Cameron Bauer.

          p.   cm.

    Includes index.

    ISBN 1-56072-528-1

    1. Algebra. I. Title.

QA152.2.B384   1998   98-35903

512—dc21   CIP

Printed in the United States of America

CONTENTS

PREFACE

This text attempts what, to the best of my knowledge, is a new approach to teaching mathematics. First and foremost, the book makes a direct appeal to students who possess a love of sports but who might not otherwise choose to study higher mathematics. Secondly, it places a greater than average emphasis on practical applications for the material.

A common technique of the text is to extract the mathematics from phenomena with which the student is likely familiar, discuss the math in the traditional terminology, and then present real-life, salary-earning applications for the math. By blending math and physics and, quite frankly, dipping rather deeply into college engineering curriculum, practical examples are found for most of the math presented in the typical algebra course. The selection of real-life examples may reveal some of the author's biases from being trained as a civil engineer. Because of the rather specialized nature of some of the disciplines discussed in the text, a more comprehensive discussion of these materials is presented in the teacher's manual.

The material in the book has been assembled rather tightly. Almost every subject area is required for subsequent sections. For example, a familiarity of centers of gravity as discussed in Chapter 5 is necessary to understand the parabolic travel of flexible objects discussed in Chapter 9. This creates a number of opportunities for discussing athletic phenomena involving quadratic equations. For this reason, it is recommended that instructors present what may seem, at the time, to be extraneous material.

The exercises in the text are unique in that they are dominated by word problems. The book contains no Jane-has-twice-as-many-apples-as-Bob problems. Rather, the problems take advantage of the relationships in sports with which the student is already familiar and ask the student to view these relationships in terms of variables.

While the athletic theme is carried throughout most of the book, Chapters 10 and 11 present the subjects of polynomial operations and logarithms in a traditional format with no athletic examples. A minor reason for this was to provide some instruction under the traditional format to avoid a "culture shock" when students continued into more advanced mathematics courses. The primary reason was that the author couldn't think of any athletic examples in these areas.

Chapter 12 presents an introduction into trigonometry. As noted earlier, the intent of the book is to use athletic examples to illustrate mathematics. While Chapters 1 through 11 were able to cover most of the material of algebra, they did not exhaust the opportunities to illustrate the math in sports. It seemed almost unfair to give students a deep familiarity of vectors and then not teach them the arctangent function to determine the angles of the motion.

My deepest gratitude is extended to my brother, Brent Bauer, for his contribution of perspectives from the aeronautical realm and for providing an overall reality check on the material.

Because of the author's orientation toward applied mathematics rather than math education, any comments and suggestions on the text would be appreciated.

Cam Bauer, P.E.

Chapter 1

WHY STUDY MATH?

One of the hardest things about learning math is understanding what it's used for. Math can often look really hard and sometimes it seems like there's no use for it. But in many ways, studying math is like training for sports.

To train for athletic competition you need to do several different things. Preparation involves lots of weight training, endurance development, agility drills, and practicing technique. A wrestler can run 10 miles a day training for a match, but during the match, he will never run at all. A large part of a football player's preparation is spent lifting weights. But on the field, a football player will never do anything that looks like a curl or a bench press. A three man weave is a common basketball drill, but no basketball player weaves down the court in a fast break. Preparing for college or a job has something in common with training for sports:

The best way to prepare for competition often looks very different from the competition itself.

It's easy to see the value of basketball drills because nearly everyone has seen a basketball game. Anyone can see how the drills help you prepare for the games. But most people have never watched someone design an airplane or do the accounting for a store. That's why it's hard to see the value of math.

WEIGHT TRAINING

Athletes in nearly every sport include some type of weight lifting in their training. This is because muscular strength is needed in almost every sport. Math is like weight training in that it improves your abilities no matter what you're training for. Almost all jobs use some kind of math, but studying math helps you in jobs in jobs that do not even use numbers. For example, when you study geometry, you learn to analyze problems and develop arguments to prove points. The kind of logic used in geometry is basically the same kind of logic lawyers use everyday.

THE 40-YARD DASH

Many times in life, large numbers of people compete for a limited number of opportunities, for example when people try out for a team or apply for a job. When this happens, the people making the selections need some type of quick test to "weed out" people. During tryouts, coaches will often have the athletes timed at the 40-yard dash. If an athlete has a good 40-yard dash time, he or she will likely make the first cut. The faster athletes will have a competitive edge over their competition.

Math skills are usually important during tests for college and jobs. Just as faster athletes have a competitive edge in sports, people with good math skills will have a competitive edge in college and in the job market. That is why the Scholastic Aptitude Tests (SATs) devote a full one-half of their test to math.

Math and science involve a great deal of competition. Individually we compete for grades in school and for positions in companies. As companies and countries we compete for prosperity. We compete with other countries to produce the best products and services. Success or failure at this competition means the difference between employment and unemployment.

HOW THE BOOK WORKS

One of the main purposes of math is to keep track of things. Among other things, accountants keep track of money; farmers keep track of crops; and engineers keep track of the forces in bridges. Math is also used to keep track of lots of things in sports. Scorekeepers keep track of points and time. Track officials monitor distances in field events. Judges give scores to gymnasts and divers. Referees use a very complicated system to keep track of position on a football field. Statisticians use math to judge competition from a number of different perspectives.

Since success in sports depends on understanding the numbers, athletes often develop some strong mathematical instincts. Athletes also learn about the motion involved in sports. All of this motion can be described using math. This book builds on the athlete's instinct for the math in sports and relates it to the math is used in jobs.

The athletic world just doesn't *have* good examples to explain mathematics, it is *full* of them. For example, the math used to calculate the average points per game is very similar to the math used to calculate the center of gravity of a football player or wrestler in a certain stance. This helps explain the value of studying math. Although math may sometimes be hard to learn, most of it can be used in lots of different ways in completely different fields. To understand math well is to open a large number of opportunities in the professional world.

One thing should be clear about this book. The book makes no effort to make you a better athlete -- *directly*. The book will, however, make you think better. *Indirectly* it will make you a better athlete.

Chapter 2

ALGEBRAIC LAWS

To start solving algebra problems, we need to learn some of the laws of algebra. Most of the laws of algebra involve some pretty simple ideas. Some of the laws are so simple, it might seem strange that there are even laws for them. Part of the purpose for algebraic laws is to help us break down complicated problems into simpler parts. By breaking down problems step by step, we can solve the most complex problems.

2.1 NUMBER LINES

Along with the laws, we're going to learn about some of the tools of math. Math uses several tools to help us visualize our work and keep track of the numbers. Math problems are always easier if we can draw pictures to help us see what's going on. The tool that's used to help us view addition and subtraction is the *number line*. The number line has negative numbers on the left and positive numbers on the right. Usually, the number line is marked with *integers*.

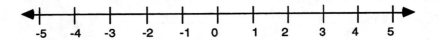

Fortunately, the number line can be compared with a similar tool used in the athletic world. The football field is marked so that players and referees can keep track of the progress of the ball down the field. The number line is similar to the yardage markers on a football field.

As you'll see several times in this book, the athletic world has lots of things which help explain algebra. Like the numbering system on the football field, the things in sports are usually more complicated and need to be simplified to explain the algebra.

The football field is numbered like this:

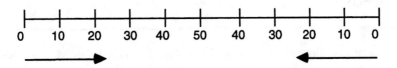

The field is marked every yard but numbered only every 10 yards. The numbers increase on both ends toward the center of the field. To show how number lines are used, most of the examples in this book will use the section of the field where the number line and the football field look alike. On the football field, this is the ascending (left) side of the field. The following examples show how the number line can be used to keep track of addition and subtraction.

Example 2.1: In a football game, the Falcons obtain possession of the ball on their 14 yard line. The following plays occur as shown below:

1st play   A 7-yard sack
2nd play   A 10-yard pass
3rd play   An incomplete pass, and
4th play   A 4-yard run

Where is the scrimmage line at the end of these plays?

Solution: This type of problem can probably be figured out in your head, but we're going to use visual tools so that we'll know how to use them with more difficult problems. To start out, we're going to put the zero point at the goal line. The progress of the ball can be displayed on a number line like this:

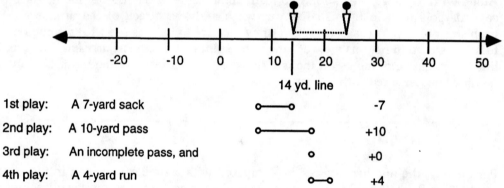

Figure 2.1 Field Position Graph

To determine the location of the scrimmage line at the end of these plays we simply add the gains and losses together. Successful runs and passes are shown as positive numbers and losses, such as sacks, are shown as negative numbers.

$$14 + (-7) + (+10) + (0) + (+4) = 21$$

The final scrimmage line is at the 21 yard line.

This example placed the zero at the goal line (on the left). When doing math problems, you can place the number line anywhere to make it best fit the problem. For example, if a team suffers a number of sacks, it will be more worried about getting a first down than a touchdown. The next problem shows an example using the first down marker as the zero point.

Example 2.2: After getting a 13-yard sack on their first down, the offense is second down and 23 yards to go. The full series of plays is shown below. Will the team get a first down?

1st play: A 13-yard sack
2nd play: A 4-yard run
3rd play: A 15-yard pass

Solution: Since we're more interested in the first down, it would be helpful to place the zero of the number line at the first down marker rather than the goal line.

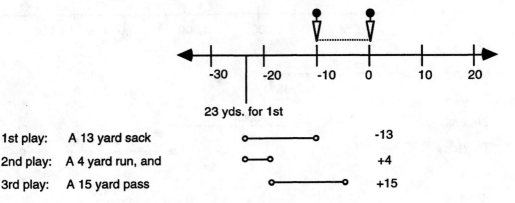

Figure 2.2 Zero Point at First Down Marker

The numbers add up as follows: (-23) + (+4) + (+15) = -4

Since we defined the zero point as the first down marker, the offense does not get a first down.

The team does not get a first down on the play (and should probably punt).

The math used to keep track of gains and losses on a football field is just like the math used by people in business everyday. Businesses have to deal with gains and losses of money. The following example shows how the number line may be use to help view the addition and subtraction involved in running a business.

<u>Example 2.3:</u> Roberta and Joe run a delicatessen. The costs of running their business include supplies, rent, staff wages, and taxes. Their income comes from the sales of their food and drinks. During the course of a week, their gains and losses occur as shown in Table 2.1. The table shows the gains as positive and the losses as negative.

<u>Day</u>	<u>Net Gain/Loss</u>
Sunday:	closed
Monday:	$976
Tuesday:	-$246
Wednesday:	$864
Thursday:	$1,019
Friday:	-$218
Saturday:	-$435

Table 2.1

Using a number line, show their cash flow over the week and determine their balance at the end of the week.

Solution: The gains and losses are added in the same way as the gains and losses in football.

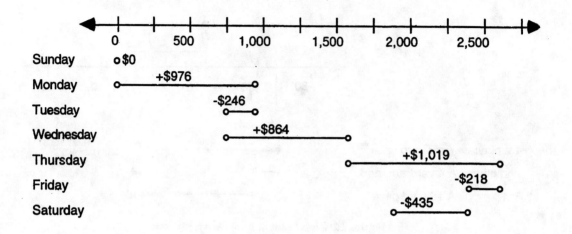

Figure 2.3 Graphing of Deli Income and Expenditures

If we add the gains and subtract the losses, the balance at the end is $1,960.

Exercise Set 2.1

Problems 1 through 4 describe the progression of the football on a series of plays. Determine the field position at the end of the series. Assume that all plays occur with the offense on the left half of the field and moving to the right. Use a number line to show the progress of the ball as done in the Examples 2.1 and 2.2.

1. Start: 14 yard line
    1st down: 4 yard run
    2nd down: 5 yard run
    3rd down: 6 yard sack
    4th down: 10 yard pass

3. Start 22 yard line
    1st down: incomplete pass
    2nd down: 6 yard pass
    3rd down: 2 yard run
    4th down: 4 yard pass

2. Start: 6 yard line
    1st down: 8 yard pass
    2nd down: 5 yard sack
    3rd down: 3 yard sack
    4th down: 4 yard run

4. Start: 12 yard line
    1st down: offsides penalty (-10 yards)
    2nd down: 12 yard pass
    3rd down: incomplete pass
    4th down: 32 yard run

The following problems relate to the deposits and withdrawals of a personal checking account. Graph the transactions on a number line as done in Example 2.3. Determine the account balance at the end of the month. Also, determine whether the account was overdrawn (had a negative balance) at any time during the month.

5.	February 1st	$647	Account balance	7.	March 1st	$1000	Account balance
	February 3rd	$335	Rent (-)		March 3rd	$335	Rent (-)
	February 5th	$770	Paycheck (+)		March 5th	$770	Paycheck (+)
	February 8th	$75	Grocery bill (-)		March 8th	$75	Grocery bill (-)
	February 15th	$297	Car payment (-)		March 15th	$297	Car payment (-)
	February 20th	$770	Paycheck (+)		March 20th	$770	Paycheck (+)
	February 22nd	$75	Grocery bill (-)		March 22nd	$75	Grocery bill (-)
	February 26th	$65	Utility bills (-)		March 26th	$65	Utility bills (-)
6.	February 1st	$647	Account balance	8.	March 1st	$0	Account balance
	February 3rd	$335	Rent (-)		March 5th	$770	Paycheck (+)
	February 5th	$770	Paycheck (+)		March 6th	$355	Rent + late fee(-)
	February 8th	$75	Grocery bill (-)		March 8th	$75	Grocery bill (-)
	February 15th	$297	Car payment (-)		March 15th	$297	Car payment (-)
	February 16th	$864	Car repair (-)		March 20th	$770	Paycheck (+)
	February 20th	$770	Paycheck (+)		March 22nd	$75	Grocery bill (-)
	February 22nd	$75	Grocery bill (-)		March 26th	$65	Utility bills (-)
	February 24th	$500	New stereo (-)				
	February 26th	$65	Utility bills (-)				

2.2 THE PROPERTIES FOR ADDITION AND SUBTRACTION

There are three main properties for addition and subtraction. Although the properties are named for addition, they also work for subtraction. This is because adding a negative number is the same thing as subtracting a positive number. Before we discuss the properties, we should first look at the concept of variables.

2.2.1 Variables

In algebra, a variable is an unknown number. The concept of variables is taught in grade schools with problems like:

$$3 + \boxed{} = 5$$

In this equation, the box is the "variable." Algebra uses letters instead of boxes. You can probably figure in your head that the answer is 2. In algebra we write the same problem like this:

Solve for x:

$$3 + x = 5$$

$$x = 2$$

Unfortunately, many of the equations we encounter in life are much more complex than this. They often look like this:

Solve for x:

$$1 = \sqrt{\frac{x^4 - 6x + 9}{x^2 + 3}}$$

And they get worse. However, by learning some of the laws of mathematics, problems like this can be broken down and solved easily.

With all of the math formulas used in the science, certain letters often get associated with certain values. For example, "d" is often used for distance and "t" is often used for time. Since there are so many different types of values and formulas in the professional world, our alphabet gets used up pretty quickly. To minimize the confusion of using the same letters for different purposes, mathematicians and scientists often use letters from the Greek alphabet for variables. The problem above could also have been written as:

Solve for α:

$$3 + \alpha = 5$$
$$\alpha = 2$$

The letter "α", or alpha, is the Greek letter for "a".

Since variables are unknown numbers, mathematicians use a special word for the known numbers. Numbers such as 3, 17, 5.26, and 1/2 are known as *constants* because their values never change. When different variables or constants are combined, they form terms known as *polynomials*. Some examples of polynomials are shown below:

$$x + 1 \qquad -x + y \qquad xy + 3 \qquad 2x + 3y - 6$$

2.2.2 The Commutative Property of Addition

A commonly used algebraic law is the Commutative Property of Addition. This property says that a set of numbers will add to the same sum regardless of the order they are added.

Example 2.4: Kyle plays in two basketball games. In the first game, he scores 10 points in the first half and 16 points in the second half. The points of the first game could be added as follows:

$$10 + 16 = 26$$

In the second game he scores 16 points in the first half and 10 points in the second half. The points of the second game could be added as follows:

$$16 + 10 = 26$$

We notice that he scored 26 points in each game. The same two numbers will add to the same total no matter what order they are added. From this example see that:

$$10 + 16 = 16 + 10$$

This is an example of the Commutative Law of Addition. In algebra, we write laws with variables instead of numbers. The precise mathematical definition is as follows:

The Commutative Property of Addition

For any real numbers a and b,

$$a + b = b + a$$

Note that the property is limited to "real" numbers. Most of the algebraic laws are limited to real numbers. An example of number which is not real is the fraction $\frac{1}{0}$. If you try to divide a number by 0 with a calculator, it will give you an "Error" message. This is because calculators can only work with real numbers. The methods of dealing with numbers that are not real are taught in calculus. All of the subjects discussed in this book are limited to real numbers.

2.2.3 The Associative Property of Addition

The Associative Property of Addition tells us we will get the same sum when we add a series of numbers regardless of how they are grouped.

Example 2.5: In a series, the Lions move the football down the field like this:

First play:	6 yard run	(+6)
Second play:	8 yard sack	(-8)
Third play:	5 yard pass	(+5)

What is their net movement at the end of the 3 plays?

Solution: The movements of the three plays can be added in several different ways. We could add the first two numbers first.

$$(6 + (-8)) + 5$$

$$= (-2) + 5$$

$$= 3$$

Or we could add the last two numbers first.

$$6 + ((-8) + 5)$$

$$= 6 + (-3)$$

$$= 3$$

At the end of the three plays, the Lions have moved the ball 3 yards forward.

The problem shows how adding the numbers gets the same result no matter how they are grouped. This is an example of the Associative Property of Addition. The formal definition of the property is as follows:

The Associative Property of Addition

For any real numbers a, b, and c,

$$(a + b) + c = a + (b + c)$$

2.2.4 The Addition Property of Equality

Another algebraic law is the Addition Property of Equality. It tells us how numbers can be added and subtracted from equations.

Example 2.6: Mike and Jane are the captains of two teams in a pickup volleyball game. Both teams naturally have the same number of players. All of the players are of relatively similar ability and the game is fair and competitive. A half hour into the game, four other players show up and ask to get in the game. Naturally, the new group of players is split and the same number of players (2) is added to each side. Since the same number of players is added to each team, the game remains fair and competitive.

In the pickup game,

m = the number of players on Mike's team
j = the number of players on Jane's team
n = the number of new players added to each team

Since Mike and Jane have the same number of players on their teams, we can write the equation:

$$m = j$$

Since they add the same number of new players, n, to their teams, we can write:

$$m + n = j + n$$

This event represents an example of the *Addition Property of Equality*. It basically says that if we have an equation, it will remain equal if we add the same amount to both sides. In mathematical terms it's stated like this:

The Addition Property of Equality

For any real numbers a, b, and c

if $a = b$, then

$a + c = b + c$

The property also holds true for subtraction; the equation will remain the equal if we subtract the same amount from both sides. In the volleyball game, the teams would have stayed equal if two players on each team had *left* the game. While this concept may seem too simple to call it a property, it is one of algebra's most important tools. The property is used over and over to solve algebra problems.

Important: Note that the first step in solving this problem was to *define the variables*. In the example, we said:

> m = the number of players on Mike's team.

Defining the variables is always necessary to remember what numbers we are trying to find. This step is always used to begin any algebra problem. Also, remember that a variable must always represent a *number*. It would make no sense to write:

> m = the players on Mike's team, or

> m = Mike's team.

The following example shows how the Addition Property of Equality is used.

Example 2.7: On the second down, the Wildcats are on their own 21 yard line. On the third down they're on their own 38 yard line. How far did they move the ball on the second down?

Solution: For this problem, we can define the terms as follows:

> s = the number yards from their goal line on the second down = 21
> t = the number yards from their goal line on the third down = 38
> m = the number yards moved on the second down

The terms are related by following
equation:

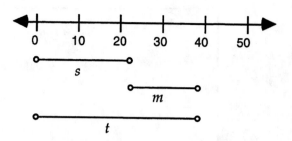

$$t = s + m$$

or $38 = 21 + m$
$$m + 21 = 38$$

Figure 2.4 Graphing of Field Position

If we want to find the value of m, we must
get the equation to look like:

$$m = \underline{\quad}$$

The key to getting the equation into this form is to get rid of the 21 on the left side of the
last equation. This is done by adding the *Additive Inverse* of +21 to each side of the
equation. The Additive Inverse of a number is the number which will cause it to add to
zero. The Additive Inverse of +21 is -21. The addition of a number to each side of the
equation is allowed by the Addition Property of Equality.

$$m + 21 + (-21) = 38 + (-21)$$

$$m + 0 = 17$$

$$m = 17 \text{ yards}$$

The Wildcats moved the ball 17 yards on the second down.

Note - In this example we used another very simple but important law of mathematics,
the *Additive Identity Property*, when we said $m + 0 = m$.

> *The Additive Identity Property*
>
> For any real number a,
>
> $$a + 0 = a$$

Exercise Set 2.2

The following problems require finding the value of a missing number. Use the procedures in Example 2.7 to determine the value of the missing number. Be sure to define the variables. Show all of your work. Unless the problem says otherwise, assume that all football plays occur on the left side of the field with the offense moving to the right.

1. On the first down, the Hawkeyes are on their 14 yard line. On the second down they're at their 43 yard line. What distance, d, did they move the ball on the first down?

2. After a 15-yard run, the Eagles are on their 27 yard line on the third down. What was there position, p, on the second down?

3. John wants to wrestle in the 150 lb. division. He now weighs 163 lb. How much weight, w, must he lose?

4. In a track meet, the Tigers are down 68 to 52. How many points, p, are they behind?

5. On her first attempt, Sharon jumped 18 feet in the long jump competition. The best jump of the day is 18 feet - 7 inches. How much farther will she need to jump to match the best jump?

6. While on the 38 yard line, the offensive team suffers a 15-yard penalty. Where is the ball placed?

7. Renee's average score on a particular golf course is 88. So far, she has 57 strokes on the course. How many more strokes can she make to meet her average score?

8. Steve needs to earn $3,000 for his college tuition next year. He needs $286 more to meet this figure. How much has he saved already?

9. In the second period of a hockey game, Larry scores 2 goals. In the third period, he scores one. For the whole game (3 periods), he scores 4 goals. How many goals did he score in the first period?

10. Individually, the first 3 runners of a 4 x 100m relay team have times of 9.8, 10.1, and 9.9 seconds. The total time of the race is 39.5 seconds. What was the individual time of the anchor?

11. During a particular game, a baseball team has three pitchers. The first pitcher allows 4 runs. The second allows 3. The opposing team scores 9 runs in the game. How many runs did the third pitcher allow?

12. A football team scores 29 points with 3 touchdowns with PATs (7 points each) and two field goals. How else did they score?

13. With only 10 seconds left in the game, an offensive team is on the defensive team's 48 yard line on second down. The comfort range of the offensive team's field goal kicker is 35 yards. During a field goal, the ball is placed 6 yards behind the scrimmage line. How far must the offensive team move the ball to get within the comfort range of the field goal kicker?

Solve for x:

14. $x - 2 = 106$

15. $4 = 16 - x$

16. $x - 2 = -14$

17. $13 + 10 = -x$

18. $(x + 5) - 3 = 27$

19. $357 = x + 1200$

20. $x + (7 - 3) = 15$

21. $x - 5{,}000 = 10{,}000 + 3{,}000$

22. $12 = x + 3$

23. $-7 + x = 23$

24. $x - 4 = 12$

25. $x + (2 \times 3) = -1$

26. $x - \dfrac{28}{4} = 6$

27. $x - 3 = 37 - 24$

28. $\dfrac{6}{2} + 7 = x - 1$

29. $x - (2 + 4) = 0$

30. $x + 0 = 14 - 12$

31. $-14 = 7 + (x + 4)$

32. $\dfrac{21}{7} - (6 - 3) = x + 4$

2.3 THE PROPERTIES FOR MULTIPLICATION AND DIVISION

There are three major properties for multiplication and division. The properties are very similar to the ones for addition and subtraction. Although the properties are named for multiplication, they also apply to division.

2.3.1 The Commutative Property of Multiplication

The Commutative Property of Addition tells us that numbers will add to the same sum regardless of the order they're added. The Commutative Property of Multiplication says that numbers will multiply to the same product regardless of the order in which they're multiplied.

<u>Example 2.8</u>: In a basketball game, June makes 2 3-point baskets and Betty makes 3 2-point baskets. Both score the 6 points. The totals can be calculated as follows:

$$2 \times 3 = 6$$

$$3 \times 2 = 6$$

Therefore $2 \times 3 = 3 \times 2$

This represents an example of the *Commutative Property of Multiplication*. The formal definition of the property is:

The Commutative Property of Multiplication

For any real numbers a and b,

$$(a)\,(b) = (b)\,(a)$$

2.3.2 The Associative Property of Multiplication

The Associative Property of Multiplication tells us that a series of numbers will multiply to the same product regardless of how they're grouped.

Example 2.9: Marcus makes 4 3-point baskets in each of 5 games. How many points did he score in 3-point baskets in these games?

Solution: The problem can be solved in several ways. The general equation is:

$$(5 \text{ games})\left(\frac{4 \text{ baskets}}{\text{game}}\right)\left(\frac{3 \text{ points}}{\text{basket}}\right)$$

We could group the first two factors:

$$((5)(4))(3)$$

$$= (20)(3)$$

$$= 60 \text{ points}$$

Or, we could group the second two factors

$$(5)((4)(3))$$

$$= (5)(12)$$

$$= 60 \text{ points}$$

Marcus scores a total of 60 points in 3-point shots in the five games.

The answer is the same regardless of how the factors are grouped. This is the point of the Associative Property of Multiplication. The formal definition of the property is:

> *The Associative Property of Multiplication*
>
> For any real numbers a, b, and c,
>
> $$(a\,b)c = a\,(bc)$$

2.3.3 The Multiplication Property of Equality

The Addition Property of Equality says an equation will remain equal if we add the same amount to both sides. There is a similar law for multiplication. It is also possible to multiply both sides of an equation by a real number and have the equation remain equal.

Example 2.10: In a baseball game, Scott batted .500 with 6 at-bats (excluding walks). How many hits did he get during the game?

Solution: In this problem, we need to find the number of hits. We'll say:

$$h = \text{the number of hits during the game}$$

The formula for batting averages is:

$$b = \frac{h}{a}$$

where a = the number of at-bats during the game (excluding walks) = 6
b = the batting average during the game = .500 = 1/2

From the equation above:

$$0.500 = \frac{h}{6}$$

$$\frac{1}{2} = \frac{h}{6} \quad \text{or}$$

$$\frac{1}{2} = \left(\frac{1}{6}\right)h$$

As with the previous examples the key to solving this problem is to isolate the unknown variable (in this problem, h). To do this we need to get rid of the 1/6 on the right side of the equation. This is done by multiplying both sides of the equation by the *Multiplicative Inverse* or *reciprocal* of the number to be eliminated. The reciprocal of a number is the number to multiply it to 1. The reciprocal of 1/6 is 6. The Multiplication Property of Equality allows us to multiply both sides of the equation by 6.

$$\frac{h}{6}(6) = (6)\frac{1}{2} = 3$$

Therefore, $h = 3$.

Scott gets 3 hits during the game.

As noted before, the Multiplication Property of Equality also works with division. This is because dividing both sides of an equation by a real number is the same as multiplying the equation by the number's reciprocal.

This property shows why the algebraic laws are limited to real numbers. If we were to divide both sides of an equation by 0, it would be the same thing as multiplying it by 1/0, which is an unreal number. Division by zero is not permitted in math.

The Multiplication Property of Equality

For any real numbers a, b, and c,

if $a = b$, then

$$ac = bc$$

Note - Again we used a very simple but important law of algebra when we said $1 \cdot h = h$. This property is known as the Multiplicative Identity Property.

The Multiplicative Identity Property

For any real number a

$$a \times 1 = a$$

Exercise Set 2.3

Show all your work. Define the variables.

1. Jane scored 16 points in 2-point field goals in a basketball game (no 3-point baskets). How many 2-point field goals did he make?

2. In a particular basketball game, Shelly's freethrow percentage was 80% (0.800). She made 4 freethrows. How many attempts did she make?

3. In a baseball game, Bill batted .400 in 5 at bats. How many hits did he get? (No walks or errors.)

4. Because of an offsides penalty near their goal line, an offensive team has the ball moved half the distance to the goal line. The ball is placed on their 3 yard line. Where was the scrimmage line when the penalty occurred?

5. Maria's paycheck shows that she earned $420 before taxes. The check covers a 80-hour pay period. How much does Maria earn per hour?

6. A barber shop receives $8 per haircut. The immediate costs of keeping the shop open, including leasing and utility costs is $96 per day. How many haircuts will the shop need to give just to cover these costs?

7. A track official is explaining the rules of a 4 x 100m relay to the runners before a race. There are 24 runners present. How many relay teams will compete in the race?

8. Annette spends $10 on buying gas at $1.25 per gallon. How many gallons of gas does she buy?

9. Over the course of a year, Brian is able to triple the amount of weight he can bench press. He ends the year benching 150 lb. How much weight could he bench at the beginning of the year?

10. A food processing plant produces cans of chili. From a batch of 231 gallons of chili, 924 cans are produced. How much chili is in each can?

Solve for x:

11. $3x = 36$

12. $\dfrac{x}{6} = 12$

13. $14 = 5x$

14. $20 = \dfrac{2}{x}$

15. $\dfrac{1}{3} = 2x$

16. $\dfrac{4}{x} = \dfrac{1}{5}$

17. $100x = 0$

18. $121 = 11x$

19. $\dfrac{3}{2x} = \dfrac{1}{4}$

20. $\dfrac{1}{5}x = \dfrac{1}{25}$

2.4 COMBINING THE PROPERTIES OF ADDITION AND MULTIPLICATION

The real strengths of the Addition and Multiplication Properties can be seen in more complicated problems where they are combined.

<u>Example 2.11:</u> The bench press bar set-up in Figure 2.5 weighs a total of 110 pounds. The bar alone weighs 10 pounds. All of the four disc weights on the bar weigh the same amount. How much does each of the four discs weigh?

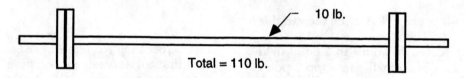

Total = 110 lb.

10 lb.

Figure 2.5 Bench Press Bar Set-up

<u>Solution:</u> Define the variable as: d = weight of each disc, in lb.

For this weight set-up:

   total weight of the set-up = weight of bar + weight of the 4 discs.

From this relationship, the following equation is formed:

$$110 = 10 + 4d$$

Again, the goal is to get the equation into the form: $d = \underline{\quad}$

We begin by getting rid of the 10 on the right side of the equation.

Using the Addition Property of Equality, $110 - 10 = 10 + 4d - 10$

$$100 = 4d$$

Now we need to get rid of the 4 on the right side of the equation.

Using the Multiplication Property of Equality, $\quad 100\left(\dfrac{1}{4}\right) = 4d\left(\dfrac{1}{4}\right)$

$$d = 25 \text{ lb.}$$

Each of the discs weighs 25 lb.

Example 2.12: In a basketball game, Jill made 5 freethrows and scored a total of 23 points (no 3-point baskets). How many field goals did she make?

<u>Solution:</u> If g = the number of 2-point field goals made,

then $\quad 23 = 2g + 5$

To solve for g, we start out with the Addition Property of Equality

$$23 + (-5) = 2g + 5 + (-5)$$

$$18 = 2g$$

Now we use the Multiplication Property of equality

$$\left(\dfrac{1}{2}\right)18 = 2g\left(\dfrac{1}{2}\right)$$

$$g = 9 \text{ fieldgoals}$$

Jill scored 9 field goals.

The formula for batting averages can be used to show how to handle polynomials when they're in the denominator.

Example 2.13: So far this season, John is batting .250 with 7 hits. He would like to get his average up to at least .300. How many consecutive hits would he need to get to have an average of .300?

<u>Solution:</u> We remember that the formula for batting averages is:

$$\text{battting average} = \frac{\text{number of base hits}}{\text{number of at - bats}} \quad \text{(excluding walks)}$$

To calculate the number of required hits, we'll need to know the number of   at-bats he had at first.

a = the original number or at-bats

$$0.250 = \frac{7}{a}$$

$$\left(\frac{a}{0.250}\right)0.250 = \frac{7}{\cancel{a}}\left(\frac{\cancel{a}}{0.250}\right)$$

$$a = 28$$

If John is to get consecutive hits, the number of additional hits will equal the number of additional at-bats.

$$c = \text{the number of consecutive hits (or at-bats)}$$

$$\text{new battting average} = \frac{\text{number of base hits} + c}{\text{number of at-bats} + c}$$

$$0.300 = \frac{3}{10} = \frac{7+c}{28+c}$$

$$\frac{3}{10}(10)(28+c) = \frac{7+c}{28+c}(10)(28+c)$$

$$3(28+c) = 10(7+c)$$

$$84 + 3c = 70 + 10c$$

$$14 = 7c$$

$$c = 2$$

John would need to get two consecutive hits to bring his batting average up to .300.

Usually, when we see a problem like this in algebra, we have to say that the variable cannot make the denominator equal zero. In this example, we would have to say that c cannot equal -28. If c equaled -28, the denominator would equal 0 and the problem would involve working with unreal numbers. The algebraic laws would not apply. But since it is impossible to get negative hits, we know that -28 cannot be a solution.

Example 2.14: In a basketball game, Tony had a free throw percentage of .667. He made a total of 8 baskets on freethrows. If he made 9 attempts during the first three quarters, how many attempts did he make in the fourth quarter?

Solution: For this problem, the terms can be defined as follows:

$$a = \text{game freethrow average} = \frac{\text{the number of successful attempts}}{\text{the number of attempts}} = 0.667$$

m = the number of successful freethrows = 8
f = the number of attempts in the first 3 quarters = 9
l = the number of attempts in the last quarter

The terms can be related by the following equation:

$$a = \frac{\text{the number of successful attempts}}{\text{the number of attempts}} \qquad \text{the number of attempts} = f + l$$

$$\text{or } a = \frac{m}{(f+l)}$$

With the numbers from the problem:

$$0.667 = \frac{8}{(l+9)} \qquad 0.667 = \frac{2}{3} \quad approximately$$

$$\left(\frac{2}{3}\right) = \left(\frac{8}{l+9}\right)$$

$$(l+9)\left(\frac{2}{3}\right) = \left(\frac{8}{l+9}\right)(l+9)$$

$$\left(\frac{2}{3}\right)(l+9) = 8$$

$$\left(\frac{3}{2}\right)\left(\frac{2}{3}\right)(l+9) = 8\left(\frac{3}{2}\right)$$

$$l+9 = 12$$

$$(l+9)-9 = 12-9$$

$$l+(9-9) = 3$$

$$l = 3 \text{ attempts}$$

Tony made 3 free throw attempts in the fourth quarter.

2.4.1 Repeating Numbers

In Example 2.14, we said that 0.667=2/3. When a basketball player has a freethrow average of 0.667 (which is said "six sixty-seven"), he or she has made 2/3 of their freethrow attempts. The number 0.667 has actually been rounded up. If you used a calculator to find the value of 2/3, you'd see it really equals 0.6666666... When a baseball player has a batting average of 0.333, he or she has hit safely 1/3 of their at-bats (excluding walks). The number 1/3 actually has a decimal value of 0.33333333.... The numbers 0.666... and 0.333... are examples of *repeating numbers*. A repeating number can also have more than one number repeating. For example the fraction 127/999 has a decimal value of 0.127127127.... In math, we shorten repeating numbers by placing a bar of the numbers which repeat. For example, the number 0.815981598159... can be written as $0.\overline{8159}$. The bar placed above the repeating numbers is known as a *vinculum*.

2.4.2 The Distributive Property of Multiplication over Addition

All of the algebraic laws so far have said that order and grouping of a calculation don't affect the answer as long as it's *all addition* or *all multiplication*. The Distributive Property of Multiplication over Addition shows how the two operations are combined. The following problem shows how the Distributive Property can be used.

Example 2.15: Wendy scores 6 baskets in the first half of a game and 5 in the second half. She shoots no freethrows or 3-point baskets. How many points does she score in the game?

Solution: This problem can be solved two ways. Our terms will be defined as:

h_1 = the number of baskets made in the first half = 6
h_2 = the number baskets made in the second half = 5
t = the total number of points scored in the game

One way to solve this problem would be to take the number of baskets made in the first half, h_1, and add them to the number of baskets made in the second half, h_2; and then take the total, $h_1 + h_2$, and multiply it by 2.

$$2(h_1 + h_2) = 2(6 + 5) = 2(11) = 22$$

Another way to solve the problem would be to take the number of points scored in the first half, $2h_1$, and add it to the number of points scored in the second half, $2h_2$.

$$2h_1 + 2h_2 = 2(6) + 2(5)$$

$$12 + 10 = 22$$

Either way, we come up with the same answer, 22.

In our example, it turned out that:

$$2(h_1 + h_2) = 2h_1 + 2h_2$$

which is what the Distributive Property of Multiplication over Addition tells us.

The Distributive Property of Multiplication over Addition

For any real numbers a, b, and c,

$$a(b + c) = ab + ac$$

The Distributive Property also applies when more than two terms are in the parentheses. For example $a(b + c + d) = ab + ac + ad$.

When worked backwards, the Distributive Property shows an important process in algebra. To say that $ab + ac = a(b + c)$ shows how the term $ab + ac$ may be *factored* into the form $a(b + c)$. The terms a and $b + c$ represent the two *factors* of the product $a(b + c)$. The ability to factor terms is necessary to solve several types of algebra problems.

In Example 2.15, the variables differentiated the values of the scores in the first and second halves by using *subscripts*. The subscripts are the small numbers on the bottom right corner of the variables. Subscripts can be either numbers or letters.

Exercise Set 2.4

Show all your work. Define the variables.

1. A curl bar set-up, similar to the bench press set-up in Example 2.11, weighs a total of 50 lb. The bar weighs 10 lb. and has a set of 10-lb. disc weights on it (no others). How many 10-lb. discs are on the bar?

2. Jennifer is batting .250 with 4 hits. How many consecutive hits will she need to get to have an average of .333?

3. A month ago, a soccer team's percent-wins was 0.500 with 12 games. They've gone undefeated since. Now their percent-win is .667 (2/3). How long was their winning streak?

4. In a wrestling match, Dan scores a total of 8 points. He scores only with 2-point and 3-point take downs. If he scored 2 3-point take downs, how many 2-point take downs did he score?

5. Sherry earned her team 9 points in a track meet where a first place earns 3 points, second place wins 2 points, and third place wins 1 point. All of her ribbons were blue or red (only firsts or seconds). If she finished first once, how many times did she finish second?

6. Cindy can save $250 a week at her summer job. She will contribute a total of $4,000 to her college tuition. So far, she has $2,500. How many more weeks does she need to work to reach the $4,000 mark?

7. A football team scores a total of 24 points with one field goal. The rest of the points were scored with touchdowns with PATs (7 points each). How many touchdowns did they score?

8. Tanya needs to buy some gasoline and a quart of motor oil. The oil costs $1 and the gas is $1.10 per gallon. If she has $12, how much gas can she buy?

Solve for x:

9. $3x - 4 = 12$

10. $28 = 6x + 3$

11. $1 + \left(\dfrac{5}{x}\right) = 9$

12. $3 = 4 + \left(\dfrac{2}{1 + x}\right)$

13. $3x + 12 = 18$

14. $\dfrac{x}{3} + 19 = 8$

15. $7x - 15 = 20$

16. $3x - \dfrac{2}{3} = -\dfrac{20}{3}$

17. $11 - 5(x - 6) = 1$

18. $4 + 3x = 60 - 5x$

19. $x + 8 = 12x + 74$

20. $3x - 113 = -86$

21. $29 - 3(x + 4) = 51 + 7(x - 6)$

22. $24 = -15 + 3\left(\dfrac{26}{x}\right)$

23. $\dfrac{5}{2}(x - 1) = 40$

24. $23 - 2x = -3$

25. $3(x - 5) + 21 = 5(x + 3) - 29$

26. $\left(\dfrac{21}{x + 4}\right) - 14 = -11$

27. $4\left(\dfrac{x + 7}{2}\right) - 13 = 31$

28. $\left(\dfrac{12}{x - 1}\right) + 2 = 0$

29. $4(x - 2) - 10 = -22$

30. $7x + 3 = 4x + 12$

31. $\dfrac{12}{x} + 4 = 16 - \dfrac{60}{x}$

32. $13(x + 7) + 3 = 29$

33. $144 = 14x + 102$

34. $\dfrac{x}{2} - 7 = 3\left(\dfrac{x}{2}\right) - 11$

2.5 Inequalities

So far in this chapter, we've tried to find a unique value of a variable. For example, a problem may have asked for the exact number of yards a football team needs to get a first down after a set of plays. But a team can get a first down by gain of more than this exact number. The best answer for that type of problem is to give a *range* of numbers which will produce a first down. To deal with this type of problem and similar problems that occur in the math, we need to understand *inequalities*. A range of values is described by using the inequality symbols. The inequality symbols are as follows:

> greater than $\geq$ greater than or equal to

< less than $\leq$ less than or equal to

$\neq$ not equal to $\approx$ almost equal to

Example 2.16: In a football game, a punt returner catches the ball in the end zone and runs it out to the 12 yard line. From this scrimmage line, what field position will earn the offense a first down? Show the range of positions on a number line which will earn the team a first down.

Solution: The offense must move the ball a minimum of 10 yards to get a first down. If they start out at the 12 yard line, they'll need to get to at least the 22 yard line for a first down. If we define x to be the offense's field position (i.e., the number of yards from the goal line), then we say that:

$$x \geq 22$$

The team must get at least to the 22 yard line to get a first down. The range is graphed as shown in Figure 2.6.

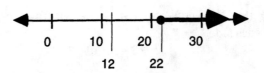

Figure 2.6 Range of First Down Gains

The solid dot at 22 indicates that the 22 yard mark is included in the range. The arrow indicates that values greater than this will also result in a first down.

To be technically correct, we can't say that they'll get a first down for any value of x greater than 22. There is also an upper limit on the number of yards gained that will result in a first down. If the team passes the goal line at the other end of the field, the result will not be a first down but rather a touchdown. To be technically correct, we would need to have upper and lower limits to the number of yards that would result in a first down. The following problem illustrates an example of a range of values with lower and upper limits.

Example 2.17: On a long bomb pass, the quarterback throws the ball into the end zone. The end zone extends 10 yards behind the goal line. If the goal line position is 0, the back of the end zone position would be -10. What is the range of values of field positions where the receiver can catch the ball which will result in a touchdown?

Solution: To score a touchdown, the ball must be caught behind the goal line but in front of the end zone. If the receiver catches the ball with his feet on the line at the back of the end zone, the pass would be ruled in complete. This line is considered to be out of bounds. The position -10 therefore cannot be in the range. To represent the fact that -10 is not included in the range, an open dot is shown at -10. Near the goal line, if the receiver catches the ball at a location where the ball has broken the plane of the goal line, the catch is good and a touchdown is scored. The location of the goal line is therefore shown as a solid dot. The range would be graphed as follows:

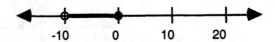

Figure 2.7 Range of Receptions Scoring Touchdowns

The range is expressed as follows:

$$0 \geq x > -10$$

Almost all of the laws in this chapter can be used with inequalities. For example, if a team gains 3 yards on their first down, the number of yards, x, on the second down that will give them a first down can be given by the inequality:

$$x + 3 \geq 10$$

You can simplify this inequality by using the Addition Property of Equality and subtracting 3 from each side.

$$x + 3 - 3 \geq 10 - 3$$

Therefore:

$$x \geq 7$$

If the team gets 7 or more yards, they'll have a first down.

The only exception to the addition and multiplication laws happens when you multiply both sides of an inequality by a negative number. When this happens, you have to reverse the inequality sign. For example, to simplify the inequality:

$$-\frac{1}{3}x \leq 12$$

We need to multiply both sides by a –3 to isolate the x. When this happens, we need to reverse the inequality sign.

$$-3\left(-\frac{1}{3}x\right) \geq -3(12)$$

$$x \geq -36$$

When graphing inequalities, it is important to remember what kind of numbers you're working with. In some problems, the only possible answer is an integer. When this is the case, non-integer numbers have to be excluded.

Example 2.18: After the eighth hole of 9-hole golf course, Debbie is shooting a 36. What is the range of additional strokes she can make and keep her total score under 41?

Solution: For this problem, we'll define the variable as follows:

x = the number of strokes on the 9th hole

Since her best possible shot is a hole-in-one, the least number of strokes she can make on the hole is 1. The lower number of the possible strokes is therefore 1. To stay under 41 strokes, she can get a total of no more than 40. The maximum number of additional strokes is therefore:

$$36 + x = 40 \quad \text{or} \quad x = 4$$

Since it is only possible to get a number of strokes that is an integer (for example, you can't get 2.7 strokes), we only count the integers. To graph this inequality, we place dots on the possible numbers instead of a solid line.

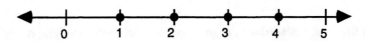

Figure 2.8 Range of Strokes

The range is therefore: $1 \leq x \leq 4$ x, an integer

2.5.1 Absolute Value

Another important tool used in describing ranges is the *absolute value*. The absolute value of a number is always the positive value of the number. For example, the absolute value of both +5 and -5 is +5. Absolute value is denoted with two vertical bars around the number. In mathematical terms:

$$|5| = 5 \quad \text{and} \quad |-5| = 5$$

The range of the section on the number line:

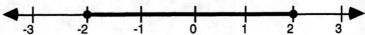

Figure 2.8 Range by Absolute Value

can be described as: $|x| \leq 2$

2.5.2 Rotation

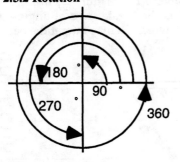

Figure 2.9

At the beginning of the chapter we saw how the number line can help us keep track of motion. Another type of motion that's common in sports and science is rotation. The most common way to track rotation is with the degree system. Figure 2.9 shows the standard 360° degree rotation system. A basketball player who does a 360° slam dunk, jumps while facing the basket, spins until he faces the basket again, and then slam dunks the ball. Rotation will be studied in much greater detail in this book.

Exercise Set 2.5

Problems 1 through 7 describe a range of values. For each problem, describe the range of values with the inequality symbols and graph the range.

1. The distance between the tee-off area and the flag on a particular golf course is 210 yards. There is a straight line of sight from the tee-off area to the hole. The putting green is a circle with a radius of 10 yards and having the hole at its center. Assuming the golfer's drive is on-line with the hole, what is the range of distances, d, of drives which will put the ball on the green.

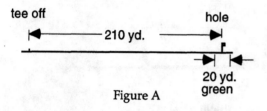

Figure A

2. A hockey goal has a height of 4 ft. With no goalie present, what is the range of heights of an on-line shot that will be good?

3. Ann has committed one foul so far in her basketball game. What is the range of additional fouls she can commit and still remain in the game? (5-foul limit)

4. As a pitcher, the least number of hits Beth has ever allowed in a game is 7. She now has a 2-hitter going. How many more hits can she allow and still break her record?

5. Jacqueline is driving to an invitational track meet. She stops for gas 150 miles from her destination. She has 3 gallons left in her 12 gallon tank. If her car gets 25 miles per gallon, what is the range of gallons she needs to buy to get her to her track meet?

6. If a ball is hit right over home plate at an angle A, as shown in Figure B, what is the range of angles, A (in degrees), where the ball will be fair?

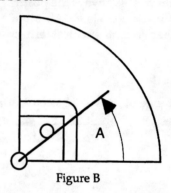

Figure B

7. A ball carrier is being held by the legs and is facing toward the backfield as shown in Figure C. He is about to be tackled. Some of his teammates are in the area. What is the legal range of angles, A (in degrees), which he may lateral the ball?

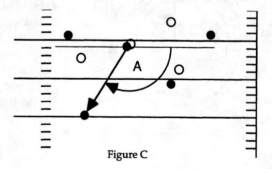

Figure C

Graph the following inequalities on a number line:

8. $12 \geq x > -2$

9. $8 > x > -1$

10. $7 \geq x \geq 0$

11. $3 > x \geq 10$

12. $5 > x \geq 3$

13. $|x| > 5$

14. $|x| \leq 10$

15. $|x| < 7$

16. $|x| \geq 16$

17. $|x - 2| > 12$

18. $|x + 2| < 8$

Simplify the following inequalities:

19. $3x > 18$

20. $x - 4 \leq 20$

21. $\dfrac{2-x}{4} > 7$

22. $7x - 3 \leq 18$

23. $5 - 3x > 20$

24. $\dfrac{x}{2} - 7 \geq 11$

Chapter 3

WEIGHT ROOM MECHANICS

One of the main reasons for studying algebra is to be able to work with the math formulas that are used in jobs. Among other things, bankers use formulas to determine the interest earned in savings accounts. Scientists use statistical formulas to determine the causes of diseases. Electrical engineers use formulas to design appliances. Architects use formulas to determine the required size of beams and columns in buildings.

This chapter explains a few of the formulas of mechanics as they can be found in the weight room. Learning to manipulate these formulas will show how the algebraic laws in the last chapter can be used. By studying the mechanics of weight lifting equipment, we can also learn some new algebraic laws.

3.1 WEIGHT LIFTING FORCES

There are basically two different types of motion involved in weight lifting. Some lifts involve a linear motion while others involve rotation. Seeing the difference between the two is important for understanding the mechanics of weight lifting.

3.1.1 Linear Forces

The squat machine helps show some of the physics involved in weight lifting equipment. In the machine in Figure 3.1, a bar slides up and down between two sets of guides. Weights are placed on the ends of the bars. In this set-up, there are generally two forces at work. Gravity pulls the weights which creates a downward force. The other force is generated by the athlete's muscles (mostly the quadriceps). To support the weight, the athlete's legs must exert a force equal to the force of the weights. The quantity of *force*, like the force generated by the muscles, can also be measured in pounds. You would say that to support 100 lb. of weights, the legs must generate 100 lb. of force in the opposite direction. To *move* the weights, the athlete's legs must exert a force *greater* than that of the weights. Again, in order to explain the algebra, we're going to make some simplifications. In all of the problems we look at here, we'll focus only on the amount of force necessary to *support* the weight rather than to move it.

Figure 3.1 The Squat Machine

3.1.2 Rotational Forces

As weight lifting equipment goes, the squat machine in Figure 3.1 is somewhat unique. Many machines do not use a simple up and down motion but use some kind of rotation instead. Figure 3.2 shows some weight lifting machines which involve rotation to generate the forces.

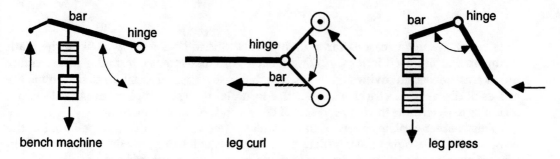

Figure 3.2 Rotating Weight Machines

In all of these machines, a bar is attached to a hinge. The athlete's muscles try to rotate the bar in one direction while the weights try to rotate it in the other direction. Muscles and bones also use rotation to generate forces. When the muscles contract, the bones rotate about their joints. Rotational forces or turning forces, like those created by weights or an athlete's muscles are known as *moments*. Moments occur in nearly all types of machinery and structures. In the mechanical world, the moment is often called *torque*.

Anyone who has ever done a curl before knows that the weight is hardest to move when it is the farthest from your body. In Figure 3.3, the weight is much harder to move at point B than it is at point A. This is because the rotational force created by the weight during a curl is the greatest at point B. The moment created by the weight tries to rotate the forearm in the clockwise direction. The purpose of curling, of course, is to develop the biceps muscle by resisting the moment created by the weight. The biceps muscle therefore tries to rotate the forearm in the counterclockwise direction. During the curl, the forearm rotates about the elbow. In mechanical terms, the point of rotation is known as the *hinge point* or *axis*. Notice that the weight always pulls the object straight toward the ground. The line that indicates the direction of the force is known as *line of force*.

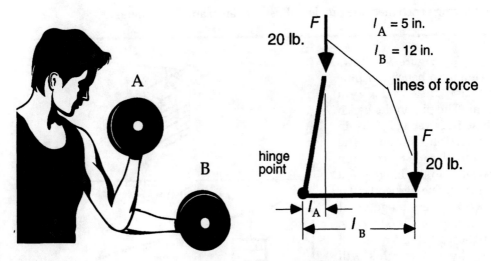

Figure 3.3 One-arm Curl

The moment, M, is calculated by multiplying two factors:

    1. The weight of the object, F (or force); and

    2. The distance between the hinge point and the line of force, l (length).

The moment, M; the force, F; and the length, l, are related by the following equation:

$$M = Fl \qquad\qquad (3\text{-}1)$$

From the diagram, the moment created by the weight at point A is:

$$M_A = Fl_A = (20 \text{ lb.})(5 \text{ in.}) = 100 \text{ in.} \cdot \text{lb.}$$

To moment at point B is:

$$M_B = Fl_B = (20 \text{ lb.})(12 \text{ in.}) = 240 \text{ in.} \cdot \text{lb.}$$

This shows that the weight is more than twice as hard to move at point B than it is at point A.

The distance used to calculate the moment is the *perpendicular* distance between the hinge point and line of force. The perpendicular distance is known as the *moment arm*. Since the moment arm is longer at point B, the moment is greater at B. This is why the weight is harder to curl at point B. A discussion is given later in the book on how to calculate moments without knowing the perpendicular distance.

The following bench machine problem illustrates an example of moments.

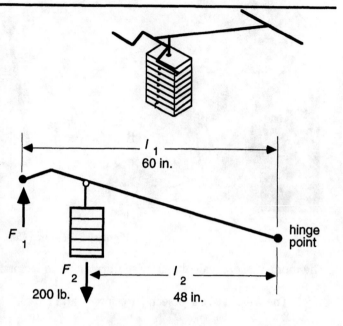

Example 3.1: Figure 3.4 shows side view of a bench machine from a Universal gym machine. The pin on the bench machine in is set at 200 lb. In the position shown, the perpendicular distance from the weights to the hinge point is 48 inches. The perpendicular distance from the hinge point to the handle bars is 60 inches. At balance, the moment created by the weights will equal the moment created by the force of the athlete's muscles. The clockwise moment created by the lifter must equal the counterclockwise moment created by the weights.

Figure 3.4 Universal Bench Press Machine

The moments of the weights and muscles are related by the equation:

$F_1 l_1 = F_2 l_2$ where

F_1 = upward force at the handle bars, in lb.
l_1 = length between handle bars and hinge point = 60 in.
F_2 = force of the weights = 200 lb.
l_2 = the length between the weights and hinge point = 48 in.

How much upward force, F_1, is required at the handles to balance the system? (The weight of the bar itself is ignored in this problem.)

Solution: To find the force necessary to balance the system, we need to solve the equation for F_1. Using the Multiplication Property of Equality, we divide both sides by l_1.

$$F_1 l_1 \left(\frac{1}{l_1} \right) = F_2 l_2 \left(\frac{1}{l_1} \right)$$

$$F_1 = F_2 l_2 \left(\frac{1}{l_1} \right)$$

Since $F_2 = 200$ lb.
               $l_2 = 48$ in.
               $l_1 = 60$ in.

Then $F_1 = \dfrac{(200 \text{ lb.})(48 \text{ in.})}{60 \text{ in.}} = 160$ lb.

An upward force of 160 lb. on the handles is necessary to balance the system.

This example helps explain one of the main reasons it is easier to lift a certain amount of weight on this type of bench machine than it is with a free bar. Since the athlete grabs the bar farther away from the hinge than the weights, the athlete has greater *leverage* over the weight. With leverage, the weights feel lighter than if you were lifting them with a free bar from directly below.

Example 3.2: Joanne is doing one-arm curls on her knee as is shown in Figure 3.5. She is using a 15-lb. weight. The figure shows how her biceps muscle attaches to the bones in her forearm (the radius and the ulna). With the distances shown in the figure, determine the force generated by her biceps muscle to balance the system when her muscle is perpendicular to her forearm.

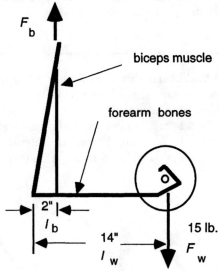

Figure 3.5 Biceps Muscle Force

Solution: As in Example 3.1, the mechanics work out to create the following relationship:

$F_b l_b = F_w l_w$ where

F_b = force on the biceps muscle, in lb.
l_b = distance between hinge point and biceps muscle = 2 in.
F_w = force of the weight = 15 lb.
l_w = distance between the weight and hinge point = 14 in.

$$F_b l_b \left(\frac{1}{l_b} \right) = F_w l_w \left(\frac{1}{l_b} \right)$$

$$F_b = F_w l_w \left(\frac{1}{l_b} \right)$$

$$= \frac{(15 \text{ lb.})(14 \text{ in.})}{2 \text{ in.}}$$

$$= 105 \text{ lb.}$$

To balance the system, Joanne's biceps muscle must generate an upward force of 105 lb.

As noted earlier, there are moments in almost any type of building or machine. The following example shows how engineers and architects use moments to design structures.

Example 3.3: A retractable backboard shown in Figure 3.6 is mounted on braces which are attached to two poles, one at each side of the backboard. The poles are attached to the wall with hinges. The backboard is suspended by a cable which allows it to be retracted when it is not in use. The backboard, braces, rim, and net weigh a total of 300 lb. The poles are 8 feet long and weigh 60 lb. each. (When you calculate moments, the weight of long cylindrical objects like poles is considered to act at their centers.) All of the moments are shown in Figure 3.6.

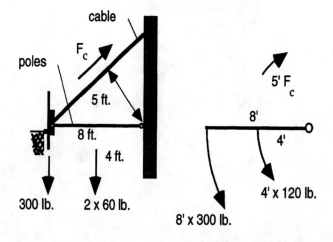

Figure 3.6 Backboard Support System

The perpendicular distance between the hinge and the cable is 5 ft. If the resisting moment of the cable equals the sum of the moment of the backboard assembly plus the moment of the poles, how much force does the cable support (i.e., how much tension is in the cable)?

Solution: In this problem, the moment results from two sources, the weight of the backboard assembly and the weight of the poles. The resisting moment of the cable will have to equal the sum of both these moments.

$F_c l_c = F_b l_b + F_p l_p$ where F_c = force on cable, in lb.

l_c = length of cable moment arm = 5 ft.

F_b = weight of backboard = 300 lb.

l_b = moment arm of backboard = 8 ft.

F_p = weight of poles = 4 x 120 lb. = 480 lb.

l_p = moment arm of poles = 4 ft.

$$F_c l_c = F_b l_b + F_p l_p$$

$$5F_c = (8 \times 300) + (4 \times 120) = 2,880 \text{ ft.lb.}$$

$$F_c = \frac{2,880}{5} = 576 \text{ lb.}$$

The force on the suspension cable will be 576 lb.

Exercise Set 3.1

The figures below show the forces and moment arm lengths on some weight lifting machines. The lengths are in inches and the weights are in pounds. Since all of the systems are at balance, the moments created by the weights are related to the moments created by the weight lifter with equation $F_w l_w = F_l l_l$. Find the missing value. Show all work.

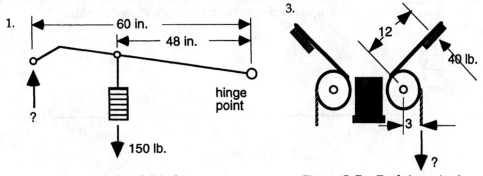

Figure A Bench Machine   Figure C Pec Deck (top view)

2.

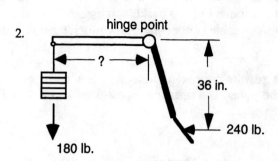

180 lb.

Figure B Leg Press

4.

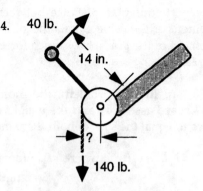

Figure D Curl Machine

5. To help protect spectators from the sun and rain, the owners of a stadium are installing an overhang. To maintain the visibility of the playing field, all of the support structures are behind the overhang as shown in Figure E. The poles which support the overhang are spaced at regular intervals around the stadium. Each pole and cable must support a section of the overhang which weighs 200,000 lb. The moment arm of the overhang has a length of 25 ft. The perpendicular distance between the cable and the hinge is 20 ft. What is the force F, on each cable?

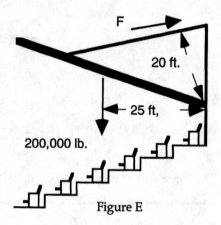

Figure E

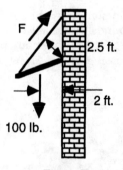

Figure F

6. A shopkeeper wishes to install a canopy above the windows and entrance to her store. The canopy will be supported by 2 cables. The canopy weighs 200 lb. Each cable will need to resist one half of the moment created by the weight of the canopy (100 lb.). The moment arm of the canopy has a length of 2 ft. The moment arm of the cables has a length of 2.5 ft. What is the force, F, on the cable?

3.2 AXIAL STRESS

The backboard problem in the last section showed how the properties of moments can be used to determine the forces on a support cable. If someone were designing the backboard set-up, the next step would be to determine how large the cable would need to be to support that amount of weight.

To keep a machine from breaking, its parts have to be large enough to support the forces that occur when it's in use. The larger the parts are, the more force they will be able to support. If the parts are too big, however, the machine will be larger and more expensive than it needs to be. To find the perfect size of the parts, engineers use a number of formulas from the field of *material mechanics*.

If you know how much force will be on a part, the required size of the part will depend on the type material it is made of. For example, most metal parts are stronger than plastic parts of the same size. Some metals are stronger than others. A type of metal or plastic can be described by how much *stress* it can safely handle. To find the perfect size for a part, engineers use formulas which determine the required size based on how much stress the material can handle. Stress is measured in units of *force per unit area*. A common unit of stress is *pounds per square inch* or *psi*.

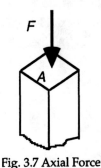

Engineers are often interested in a type of stress known as *axial stress*. The stress is called "axial" because the stress occurs along the axis of a bar or rod. The concept of axial stress is shown in Figure 3.7. The F represents the force of weight on the rod. Effectively, a weight is being placed on top of the rod which causes it to be pressed together. Two factors will determine how much weight the rod can support:

1. The cross-sectional area, A, of the rod; and

Fig. 3.7 Axial Force

2. The amount of stress, S, the material can safely handle.

The cross-section of a part is the surface you would get if you sliced it perpendicular to its axis. The cross-sectional area is the area of this surface. The cross-sections of two rods are shown in Figure 3.8.

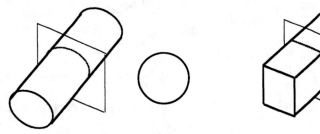

Figure 3.8 Cross-sections

The formula which relates force, area, and axial stress is:

$$S = \frac{F}{A}$$

where S = stress   (3-2)

F = point force

$F = SA$ A = cross-sectional area

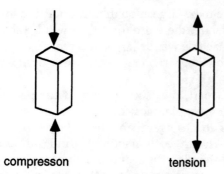

The formula holds true whether the bar is being pressed together, like a column in a building, or being pulled apart like the cable in the backboard example. When a bar is being pressed together, it is said to be in *compression*; when it is being pulled apart, it is said to be in *tension*.

compresson   tension

Figure 3.9 Axial Stresses

Example 3.4: The rod shown in Figure 3.10 is expected to carry a maximum axial load, F, of 5,000 lb. The metal alloy of the rod can safely handle a stress of 20,000 lb./in^2 (psi). What   should the cross-sectional area, A, of the rod be?

Solution: To determine the required area, we solve Equation 3-2 for A.

Multiplying both sides by A, $SA = \left(\dfrac{F}{A}\right)A$

$$SA = F$$

Dividing both sides by S:

$$SA\left(\frac{1}{S}\right) = F\left(\frac{1}{S}\right)$$

$$A = \frac{F}{S} = \frac{5{,}000 \text{ lb.}}{20{,}000 \, {}^{\text{lb.}}\!/_{\text{in.}^2}}$$

$$A = \frac{1}{4} = 0.25 \text{ in.}^2$$

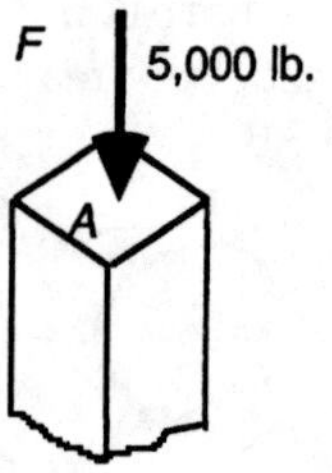

F ↓ 5,000 lb.

Fig. 3.10 Rod in Compression

The rod needs a cross sectional area of 0.25 square inches to safely support a 5,000 lb. load.

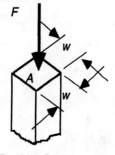

Fig. 3.11 Bar Dimensions

In the last example, we determined the required area of a square rod. To manufacture a rod like this, you would need to know more than just the required area of the cross-section. You would need to know the dimensions or width of the rod. The problem stated that the rod was square in cross-section. If the rod has a width of length w, the rod will have an area of:

$$A = w \times w = w^2$$

Since it represents the area of a square, the term w^2 is said "w squared." The "2" in the term is known as the *exponent*.

The problem required that the area equal 0.25 square inches. To find the value of w, we need to solve the following equation for w:

$$w^2 = 0.25$$

None of the properties we've learned so far have shown how to solve this kind of equation. To find the value of w, we need to take the *square root* of 0.25. This is done by finding a number which, when multiplied by itself, equals 0.25. Taking the square root of both sides of the equation,

$$A = w^2 = 0.25$$

$$\sqrt{w^2} = \pm\sqrt{0.25}$$

We need to say that w equals *plus or minus* the square root of 0.25 because both two negative and two positive numbers will multiply to a positive number. For example:

$$3 \times 3 = 9 \quad \text{and} \quad (-3) \times (-3) = 9$$

When we take the square root of a number, we have to say it may be either positive or negative. Therefore:

$$\sqrt{w^2} = \pm\sqrt{0.25}$$
$$w = \pm 0.5 \text{ in. or } \frac{1}{2}\text{in.}$$

Since we know that the width of the rod can only be positive, we can say that the rod needs to be 0.5 in. in width.

3.2.1 Area

The concept of area is an important one used in lots of different fields. Some equations for areas of other shapes are shown in Figure 3.12.

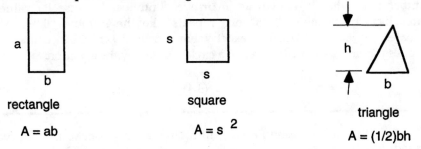

rectangle
$$A = ab$$

square
$$A = s^2$$

triangle
$$A = (1/2)bh$$

Figure 3.12 Equations for Area

The area of a rectangle is the height times the width. A square is a rectangle whose height equals its width. The area of a square is therefore the square of the side. The area of a triangle is one-half the height times the width.

3.2.2 Circles

To find the area of a circle, you need to use a special equation. To use the equation, you have to understand the concept of π or *pi*. Pi (pronounced "pie") is the Greek letter for "p." In math it represents the ratio of the circumference of a circle to its diameter.

If you cut a piece of string so that it wrapped perfectly around a circle and then stretched it out as shown in Figure 3.13 the string would have a length of π times the diameter of the circle. The length of the string equals the *perimeter* or *circumference* of the circle. The circumference, *c*, of a circle is equal to π*d* or 2π*r*. In mathematical terms:

$$c = \pi d = 2\pi r \qquad \text{where} \quad c = \text{the circumference of the circle} \qquad (3\text{-}3)$$
$$d = \text{the diameter of the circle}$$
$$r = \text{the radius of the circle}$$

The relationship holds true for any size of circle.

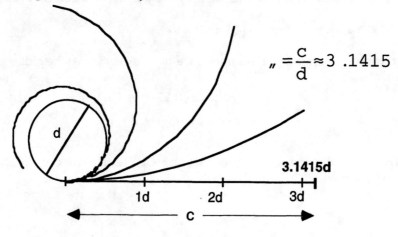

$$" = \frac{c}{d} \approx 3.1415$$

Figure 3.13 Pi

The number π is what is known as an *irrational* number because its value neither terminates like the number "1.25" nor repeats like the number "0.333. . .". It is approximately equal to 3.1415. The exact value of π is described in courses teaching infinite series (not in this text). The area of a circle is given by the equation:

$$A = \pi r^2 \qquad\qquad (3\text{-}4)$$

<u>Example 3.5:</u> If the rod in Example 3.4 had a circular cross-section, what would the required radius of the bar be?

<u>Solution:</u> In Example 3.4 the required area was 0.25 in.^{2}. The required radius of the bar would be *r*, where:

$$A = \pi r^2 = 0.25 \text{ in.}^2$$

$$r^2 = \frac{0.25}{\pi}$$

$$r = \pm\sqrt{\frac{0.25}{\pi}}$$

Again, since we know the radius is a positive number, we can ignore the negative value. Therefore:

$$r = \sqrt{\frac{0.25}{3.1415}} = \sqrt{0.0796} = 0.28 \text{ in.}$$

The required radius of a bar with circular cross section would be 0.28 in.

The quantity of stress has the same units as the quantity of *pressure*. You may have recognized the units of pounds per square inch (psi) as a unit of air pressure. Pressure has the same relationship with force and area that stress does. The following problem shows an example of air pressure.

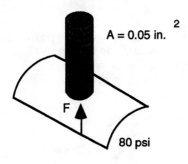

$A = 0.05 \text{ in.}^2$

F

80 psi

Figure 3.14 Air Pressure

Example 3.6: A bicycle tire is inflated to 80 lb./in.^{2}. The valve has an area of 0.05 in.^{2}. What is the force on the valve?

Solution: In this problem, we substitute the value of pressure for stress into Equation 3-2 and solve for F, the point force.

$$F = SA$$

$$= \left(80 \text{ }^{lb.}\!/_{in.^2}\right)\left(0.05 \text{ in.}^2\right)$$

$$= 4 \text{ lb.}$$

The valve has a force of 4 lb. on it.

Exercise Set 3.2

1. An engineer is designing a machine and is arranging the calculations of the axial stress of several parts into a table as shown below. Using Eq. 3.2, complete the table.

Part	Area, A (in.^{2})	Force, F (lb.)	Stress, S (lb./in.^{2})
A	2	30,000	
B		10,000	20,000
C	6	12,000	
D	0.5		1,000
E		12,500	5,000

Table A

2. After doing her stress calculations, an architect has determined that a square column will need a cross-sectional area of 64 in.^{2}. What is the required width, w, of the column?

3. To handle an axial load, a bar with a rectangular cross-section in a machine needs a cross-sectional area of 10 in.^{2}. To fit in the machine, it needs a thickness of 2.5 in. What is its width, w?

4. Sam is painting one side of a very long fence. The fence is 6 ft. tall. He has enough paint to cover an area of 1,200 square feet. How long of a section of fence will he be able to paint?

5. As a result of a design change, a rod with a rectangular cross-section must be changed to have a square cross-section. The original rectangle had dimensions of a and b. The square has a width of w. Since the rod must support the same force, the area of the rod must remain the same. Determine the value of w in terms of a and b.

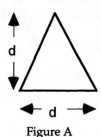

Figure A

6. Teri has a triangular patch in her yard where she would like to plant a lawn as shown in Figure A. The triangle has a base width of d and a length of d. Since lawn seed is sold by the amount of area it will cover, how large of an area, in terms of d, should she buy seed for?

7. A circle fits in a square as shown in Figure B. The square has sides of length s. What is the area of the circle?

Figure B

Figure C

8. A painter is painting the lines on a gym floor. The jump circle at mid-court is a circle with a line through its center. The line has a length of x. The painter charges by the length of the

striping. What is the length of striping, in terms of x, needed to paint the jump circle?

9. In a 4 x 400 m relay race, two sprinters receive their batons at the same time at the finish line as shown in Figure D. The runners must stay in their lanes and reach the finish line to complete the race. Given the dimensions of the track, how much farther must the sprinter in lane 2 run to tie with the sprinter in lane 1?

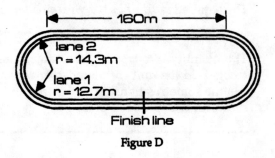

Figure D

10. An overhang above a patio is supported by two rods as shown in Figure E. The overhang has a weight of 500 lb. and moment arm length of 10 ft.

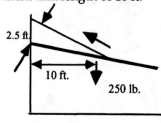

Figure E

Each rod must resist the moment created by half of the overhang (250 lb.). The moment arm of the rods is 2.5 ft. The rods are circular in cross-section and are made of a metal that can safely handle 20,000 pounds per square inch. What is the required diameter of the rods?

3.3 BENDING STRESS

Another type of stress you find in the weightroom is *bending stress*. When a bar experiences a moment, it will bend. Figure 3.15 shows some exaggerated bending of a bar on one side of a bench press.

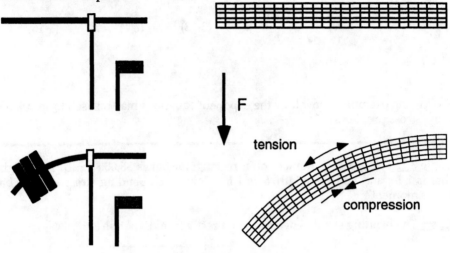

Figure 3.15 Bending Stress

For the bar in Figure 3.15, the bending creates a stress which makes the top of the bar pull apart. The bottom part of the bar will be in compression. The stress will be the greatest at the top and bottom edges of the bar. To determine the best size for a bar that will experience bending, engineers use formulas for bending stress. The variable that's usually used for bending stress is sigma, σ, the Greek letter for "s". The formula for the bending stress of a bar with a circular cross-section (like on a bench press) is given by the equation:

$$\sigma = \frac{4M}{\pi r^3}$$ where   σ = the bending stress on the bar   (3-5)

M = the moment on the bar
r = the radius of the bar

The formula for the bending stress of a bar with a rectangular cross-section is given by the equation:

$$\sigma = \frac{6M}{bh^2}$$ where   σ = the bending stress on the bar   (3-6)
M = the moment on the bar
h = the height of the bar
b = the base width of the bar

3.3.1 Properties of Exponents

As you can see, the equations for bending stress get rather complicated. You can find formulas with complicated exponents in many types of jobs. To be able to work with them, you need to understand the properties of exponents. The equations below show the laws for exponents:

$$a^m \cdot a^n = a^{m+n} \quad (3\text{-}7)$$

$$a^{-n} = \frac{1}{a^n} \quad (3\text{-}11)$$

$$\sqrt[n]{a} = a^{\frac{1}{n}} \quad (3\text{-}14)$$

$$\sqrt[m]{\frac{a}{b}} = \frac{\sqrt[m]{a}}{\sqrt[m]{b}} \quad (3\text{-}8)$$

$$a^1 = a \quad (3\text{-}12)$$

$$a^0 = 1 \quad (a \neq 0) \quad (3\text{-}15)$$

$$\sqrt[n]{a^m} = \left(\sqrt[n]{a}\right)^m \quad (3\text{-}9)$$

$$\frac{a^m}{a^n} = a^{m-n} \quad (3\text{-}13)$$

$$(a^m)^n = a^{mn} \quad (3\text{-}16)$$

$$\left(\frac{a}{b}\right)^m = \frac{a^m}{b^m} \quad (3\text{-}10)$$

The following examples show how the exponent equations may be used to solve some bending stress problems.

Example 3.7: The greatest moment on a rectangular bar is 36,000 in. lb. The height, h, of the bar is 2 in. and its base width, b, is 1 in. How much bending stress will the bar have with this moment?

Solution: The bending stress is determined directly from Equation 3-6.

$$\sigma = \frac{6M}{bh^2}$$

$$= \frac{6(36,000 \text{ in. lb.})}{1 \text{ in.} (2 \text{ in.})^2} = 54,000 \text{ lb.}/\text{in.}^2$$

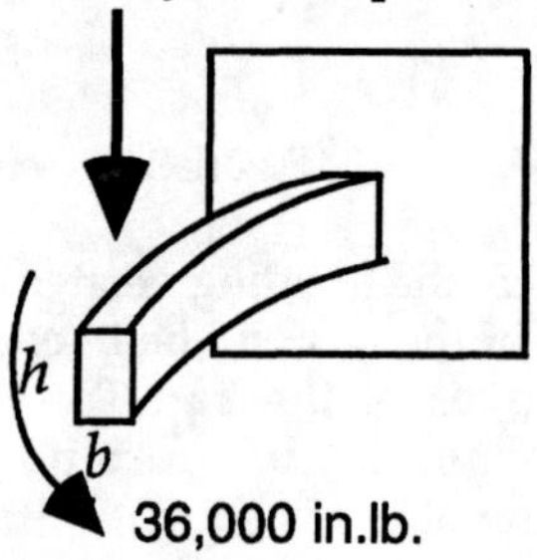

36,000 in.lb.

Figure 3.16 Bar in Ex. 3.7

The bar will experience 54,000 psi of bending stress.

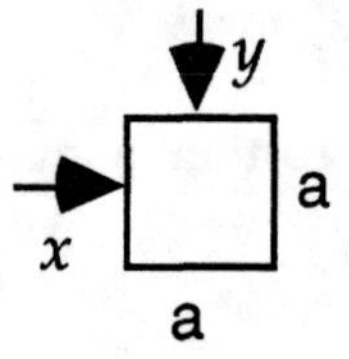

Fig 3.17 Bar in Ex. 3.8

Example 3.8: Figure 3.17 shows the cross-section of a square bar. A manufacturer is designing the bar which can bend in either direction-x or direction-y as shown in the figure. To resist bending in both directions, the bar has been given a square cross-section. Using Eq. 3-6, derive a formula that will allow you to calculate the required width, a, of a bar with a square cross-section for any values of M and σ. In other words, solve Eq. 3-6 for a if $a = b = h$.

Solution: If $a = b = h$, Eq. 3-6 could be written as:

$$\sigma = \frac{6M}{bh^2} = \frac{6M}{aa^2}$$

To solve the equation for a, we can first combine the a's. Using Eq. 3.7:

$$aa^2 = a^1 a^2 = a^{1+2} = a^3$$

Therefore:

$$\sigma = \frac{6M}{aa^2} = \frac{6M}{a^3}$$

We then try to get the a^3 term on one side. We do this by dividing both sides of the equation by σ and multiplying both sides by a^3.

$$\sigma\left(\frac{a^3}{\sigma}\right) = \frac{6M}{a^3}\left(\frac{a^3}{\sigma}\right) \qquad \text{for } a \neq 0 \text{ and } \sigma \neq 0$$

$$a^3 = \frac{6M}{\sigma}$$

Solving for a means getting its exponent to 1. This can be done by taking the *third root* of each side of the equation.

$$\left(a^3\right)^{\frac{1}{3}} = \left(\frac{6M}{\sigma}\right)^{\frac{1}{3}}$$

From Eq. 3-16: $\qquad \left(a^3\right)^{\frac{1}{3}} = a^{3 \cdot \frac{1}{3}} = a^1$

From Eq. 3-12: $\qquad a^1 = a$

Therefore: $\qquad a = \left(\frac{6M}{\sigma}\right)^{\frac{1}{3}}$

Using Eq. 3-14, we can get the equation into a more standard from:

$$\left(\frac{6M}{\sigma}\right)^{\frac{1}{3}} = \sqrt[3]{\frac{6M}{\sigma}}$$

The value of a is determined by the equation: $a = \sqrt[3]{\frac{6M}{\sigma}}$

Example 3.9: The equation for bending stress of a bar with a rectangular cross-section, $\sigma = \frac{6M}{bh^2}$, was derived from a general equation for bending stress, $\sigma = \frac{Mc}{I}$, and two equations which apply specifically to rectangular cross-sections. The two equations are:

$$c = \frac{h}{2} \qquad \text{and} \qquad I = \frac{bh^3}{12}$$

Using the three equations and the laws of exponents, derive Eq. 3-6.

Solution: $\qquad \sigma = \frac{Mc}{I} \qquad c = \frac{h}{2} \qquad I = \frac{bh^3}{12}$

We first substitute the equations for c and I:

$$\sigma = \frac{Mc}{I} = \frac{M\left(\dfrac{h}{2}\right)}{\dfrac{1}{12}bh^3} = \frac{12Mh}{2bh^3}$$

$$= \frac{6Mh}{bh^3}$$

The h's are all moved to the denominator by using Eq. 3-11:

$$\sigma = \frac{6Mh}{bh^3} = \frac{6M}{bh^{-1}h^3}$$

Eq. 3-13 is then used to reduce the h's:

$$\frac{6M}{bh^{-1}h^3} = \frac{6M}{b(h^{3-1})} = \frac{6M}{bh^2}$$

Therefore:

$$\sigma = \frac{6M}{bh^2}$$

Exercise Set 3.3

1. A bar has a base width of 1 in. and a height of 4 in. ($b = 1$, $h = 4$)
a.) If a force creates a moment of 8,000 in·lb on the bar, what is its bending stress?
b.) If the bar is placed on its side (base width = 4 in., height = 1 in.) and then has the same moment, what will its bending stress be?

2. A rod with a circular cross-section experiences a moment of 12,000 in.·lb. The rod is made of a material that can safely handle 20,000 pounds per square inch of stress. What should the radius of the bar be?

3. A rectangular bar which resists a bending moment is to be replaced with a circular bar. The bar will be made of the same material and will experience the same moment. It will therefore be sized to have the same amount of stress. If the maximum stress in the bars remains the same, the dimensions of the bars will be related by the equation:

$$\sigma = \frac{4M}{\pi r^3} = \frac{6M}{bh^2}$$

Given the equation, determine the value of r in terms of the dimensions of the rectangular bar (i.e. solve for r in terms of b and h.)

4. An engineer wants to determine how large of a moment a beam can handle with different dimensions. Solve Eq. 3-6 for M.

5. Two bars with rectangular cross-sections are made of the same material ($\sigma_{max} = 20{,}000$ psi) and have the same dimensions, $b = 2$ in. and $h = 4$ in.
a.) If the bars are fastened together as shown in Figure A, how large a moment can they support? (Use the result of problem 4.)
b.) If they are fastened together as in Figure B, how large of a moment can they support?

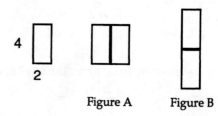

Figure A   Figure B

6. A wood beam has a base width of 1 in. and a height of 10 in. The wood can safely handle a stress of 6,000 psi. If the wood beam is to be replaced with a steel beam (σ_{max} = 30,000 psi) with the same base width, what height must it have to resist the same moment?

Solve for x:

7. $x^3 = 27$

8. $\dfrac{x^4}{x^2} = 9$

9. $x^2 = 49$

10. $x^3 - 21 = -13$

11. $x = (25)^{\frac{1}{2}}$

12. $x^2 x^2 = 16$

13. $\dfrac{x^{-3}}{x^{-5}} = 0.16$

14. $x = \sqrt[3]{64}$

15. $x^3 = 0.064$

16. $x^5 = 81\left(x^3\right)$

17. $x^3\left(x^{-2}\right) = 7$

18. $\left(x^{\frac{2}{5}}\right)^{\frac{5}{2}} = 27$

19. $\sqrt[3]{\dfrac{x^3}{8}} = 1$

20. $5^x = 1$

21. $\dfrac{1}{5^{-3}} = 25x$

22. $\left(\sqrt[4]{7}\right)^x = \sqrt[4]{7}$

23. $\left(3^2\right)^3 = 6x$

24. $\dfrac{1}{\sqrt[3]{x}} = 27$

Simplify:

25. $\dfrac{x^4 y^{-3}}{x^2 y^5}$

26. $\sqrt[3]{x^2}\left(\sqrt[3]{x}\right)$

27. $\dfrac{1}{x^{-7}}$

28. $\left(\dfrac{x^2}{y^3}\right)^4$

29. $\sqrt{\dfrac{x^4}{y^6}}$

30. $\left(x^4 y^{-2}\right)\left(x^2 y^5\right)$

3.4 DESIGNING WEIGHTS

Athletes usually prefer their weights to have even increments like 5 lb., 10 lb., and 25 lb. so that it is easy to keep track of how much they're lifting. This section shows how weights of certain amounts can be designed. Since weights are also used in business and science, this section has some practical uses. More importantly, it shows one of the main uses of math in general. You could make a 10-lb. weight in two ways. You could start out with a very large object and keep shaving it down until it weighed exactly 10 lb. or you could use the laws of math and physics to *predict* how large it needs to be.

3.4.1 Force

One of the most important things to understand about mechanics in the weight room is the concept of force and its relationships with mass and acceleration. The force created by the weight is actually the product of two factors:

1. The *mass* of the object; and

2. The pull or *acceleration* of gravity

In math terms, a force, like the force created by a weight, can be determined by the following equation:

$$F = ma \qquad \text{where} \quad \begin{aligned} F &= \text{force} \\ m &= \text{mass} \\ a &= \text{acceleration} \end{aligned}$$

Scientists refer to this equation as *Newton's first law* (after its discoverer Isaac Newton). There are lots of examples in sports that help explain this equation. In the weight room, if you bench 20 lb. of weight with the same force you bench 100 lb., the weight will move much faster. Since the mass is less and you're generating the same amount of force as for 100 lb., the 20 lb. *accelerates* faster.

The gravitational pull or acceleration on an object depends on the size of the object pulling it. Any two objects will have some gravitational pull between them. It takes an object the size of a planet to create enough gravity to be really noticeable. If you took three identical weights and attached them to three identical springs and then held one on the moon, one on Earth, and one on the planet Jupiter (which is much larger than Earth), the weights would stretch the spring as shown in Figure 3.18.

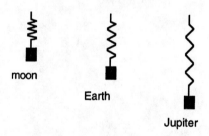

Figure 3.18 Gravitational Forces

Since the three planets have different sizes, their gravitational pulls are different and therefore stretch the spring differently. The larger the planet, the stronger the pull.
Gravity is not the only form of acceleration. Whenever something causes an object to move or change speed, we say the object is accelerating. For example, whenever a ball is hit or kicked, it accelerates.

It is important to understand the difference between mass and weight. Weight represents a force. It is the combination of mass and acceleration. The U.S. unit of the pound is a unit of force.

Every object has some mass. An object will only have weight when it is near another object and is pulled by its gravitational acceleration. The metric unit of the kilogram is a unit of mass.

3.4.2 Volume

The weight of an object can be determined by its size or *volume* and the type of material it is made of or its *specific weight*. The volume of a box, as shown in Figure 3.19, is the product of its three dimensions. The box has a volume determined by the equation:

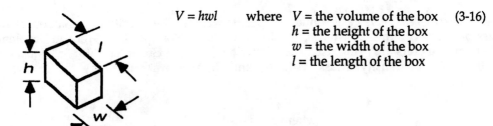

$$V = hwl \qquad \text{where} \quad V = \text{the volume of the box} \qquad (3\text{-}16)$$
$$h = \text{the height of the box}$$
$$w = \text{the width of the box}$$
$$l = \text{the length of the box}$$

Figure 3.19 Dimensions of Box

A *cube* is a box whose sides all have the same length. If a cube had sides of length s, the volume of the box would be $s \times s \times s$ or s^3. Since it represents the volume of a cube with sides of length s, the term s^3 is said "s cubed".

Fig. 3.20 Cube

Since each of the dimensions of a box has a unit of length, their product (the volume) has the unit of length to the third power or length cubed. The units of cubic inches, cubic feet, and cubic meters are examples of units of volume. The volume of liquids like milk or gasoline are often measured in units of gallons, pints, or liters.

Another type of shape which occurs often is the *cylinder*. When people think of cylinders, they usually think of circular cylinders. A cylinder is actually any object that has a constant cross-section along its length. Different types of cylinders are shown in Figure 3.21. The volume of a cylinder can be calculated by multiplying the area of its cross-section, A, by its length, l. The volume is given by the equation:

$$V = Al. \qquad (3\text{-}17)$$

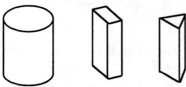

Figure 3.21 Cylinders

Remember that the units of area are always length squared. With area times length, the units of the volume of a cylinder remain length cubed.

$$Al = (l \times l)l = l^3$$

3.4.3 Specific Weight and Density

The *specific weight* of a material is defined as its *weight per unit volume*. A block of steel is much heavier than a block of wood the same size. This is because steel is a denser material than wood. The letter γ, or gamma, the lower case Greek letter for "g", is usually used for specific weight. Some examples of the specific weights of materials include steel, which has a specific weight of about 0.28 lb./in.^{3}, and water, which has a specific weight of about 0.036 lb./in.^{3}. Weight, volume, and specific weight are related by the following equations:

$$\gamma = \frac{w}{V} \qquad \text{where } \gamma = \text{specific weight}$$

$$\text{or} \quad w = \gamma V \qquad \qquad w = \text{weight} \qquad (3\text{-}18)$$

$$V = \text{volume}$$

Another term, *density*, is very similar to specific weight. The density of a material is the *mass* per unit volume. Density is usually used instead of specific weight in problems where metric units are being used. The letter ρ or *rho*, the Greek letter for "r," is usually used to denote density. Mass, volume, and density are related by the following equations:

$$\rho = \frac{m}{V} \qquad \text{where } \rho = \text{density}$$

$$\text{or} \quad m = \rho V \qquad\qquad m = \text{mass} \qquad\qquad (3\text{-}19)$$

$$V = \text{volume}$$

Example 3.10: A manufacturer of weight lifting machinery needs to produce 10 lb. weights for the stacks in its equipment. It will cut them from 1-in. thick plates of cast iron. Cast iron has a specific weight of 0.25 lb./in.^{3} (i.e., $\gamma = 0.25$ lb./in.^{3}). If the manufacturer wants the plates to be 4 in. wide, how long do they need to be to weigh 10 lb.?

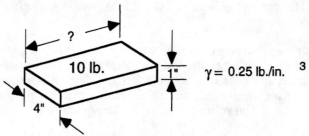

Figure 3.22 Stack Weight

Solution: Since volume of a the box can be given by the equation $V = hwl$ and $W = \gamma V$, the weight of the box can be formed with the equation:

$$W = \gamma V = \gamma hwl$$

In this problem, we know W, γ, h, and w. The problem asks for l. To determine the value of l, we solve the equation for l.

$$W = \gamma hwl$$

$$l = \frac{W}{\gamma hw}$$

$$= \frac{10 \text{ lb.}}{\left(0.25 \, \frac{\text{lb.}}{\text{in.}^3}\right)(1 \text{ in.})(4 \text{ in.})}$$

$$= 10 \text{ in.}$$

With a 1-in. thickness and a 4-in. width, the plate needs to be 10 in. long to weigh 10 lb.

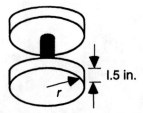

Fig. 3.23 One-hand Barbell

Example 3.11: A manufacturer of one-hand barbells needs discs weighing 21 lb. The discs are welded together to make 45 lb. weights. The handle between the discs weighs 3 lb. (2 x 21 lb. + 3 lb. = 45 lb.) The discs will be cut from sheets of 1.5-in. thick sheets of metal with a specific weight of 0.3 lb./in.^{3}. What should the radius, r, of the discs be?

1.5 in.

r

Solution: Again, the weight of the disc is its volume times its specific weight. Each of the discs is a thin cylinder. The volume equals the area of its cross-section times its length.

$$W = \gamma A l \qquad \text{where } A = \pi r^2$$

$$= \gamma\left(\pi r^2\right)l$$

We need to solve this equation for r.

$$W = \gamma \pi r^2 l$$

$$r^2 = \frac{W}{\gamma \pi l}$$

$$r = \pm\sqrt{\frac{W}{\gamma \pi l}}$$

$$= \sqrt{\frac{21 \text{ lb.}}{\left(0.3 \ {}^{\text{lb.}}\!/_{\text{in.}^3}\right)\pi(1.5 \text{ in.})}} = 3.9 \text{ in.}$$

The discs must have a radius of 3.9 in. to weigh 2 lb. each.

As you may have noticed, the rectangular plate in Example 3.10 isn't much good without any holes in it. The plate will need holes for the rods which keep the plates straight or lift them. A hole is also needed for the pin. If we drilled the holes after sizing the plate to 10 lb., it would no longer weigh exactly 10 lb. To make it weigh the exact amount, we need to take the volume of the holes into consideration before sizing the plate. We can do this by subtracting the volume of the holes from the volume of the plate. The shape of the holes is a circular cylinder. The following problem shows how to compensate for the holes while designing a weight.

Example 3.12: A manufacturer wants to make 25-lb. disc weights for a set of free weights. The weight will be made from 1-in. thick sheets of steel having a specific weight of 0.3 lb./in.^{3}. One-inch diameter holes will be drilled in the discs for the bar. What should the radius, R, of the disc be?

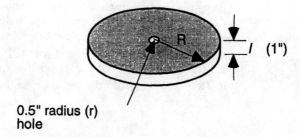

0.5" radius (r)
hole

Figure 3.22 Disc Weight

Solution: For the disc to weigh exactly 25 lb., we need to take into consideration the volume of the hole. Both the disc and the hole are cylinders. Generally, the different volumes can be related by this equation:

final volume = (volume of disc) - (volume of hole)

In the figure,

$$W = \gamma V \qquad\qquad V = Al$$
$$= \gamma(A_{disc} - A_{hole})l$$
$$= \gamma(\pi R^2 - \pi r^2)l$$

To determine the radius, we need to solve this equation for R.

$$W = \gamma\pi(R^2 - r^2)(l)$$

$$(R^2 - r^2) = \frac{W}{\gamma\pi l}$$

$$R^2 = \frac{W}{\gamma\pi l} + r^2$$

$$R = \sqrt{\frac{W}{\gamma\pi l} + r^2}$$

$$= \sqrt{\frac{25}{(0.3)\pi(1)} + 0.5^2} = 5.2 \text{ in.}$$

The discs should have a diameter of 5.17 in.

Exercise Set 3.4

The following problems ask for some missing dimensions of some objects that weigh a certain amount. It is recommended that you complete the algebra before you begin substituting the numbers into the equations.

1. A manufacturer of elevators makes 100-lb. weights which are used as counter balances. The weights will be cut from 2-in. thick sheets of steel and will have two 1-in. diameter holes in them. To fit in with the rest of the elevator equipment, the plates will have a width of 8 in. The plates will be made from a steel with a specific weight of 0.3 lb./in.^{3} How long must the plates be?

2. A manufacturer of weight lifting equipment wants to produce a 15-lb. bar from 1-in. diameter steel with a specific weight of 0.28 lb./in.^{3}. How long do the bars need to be?

3. A large cylindrical tank in a chemical plant is 10 ft. tall and has a radius of 4 ft.
a.) What is the volume of the tank?
b.) If the tank is filled with a fluid that weighs 60 lb./ft.^{3}, what is the weight of the fluid?

4. The weights in a stack of a particular type of weight machine weigh 10 lb each. To help lifters transition between the 10-lb weights, the manufacturer wants to produce a 5-lb weight that can be placed on top of the stack. To be able to slide on and off the top of the stack, the weights will have a 1-in. wide groove in them that is 2 in. long as shown in Figure A.

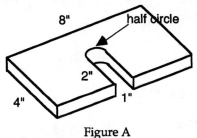

Figure A

At the end of the groove is a half circle with a diameter equal to the width of the groove. To fit on the stack, the outside dimensions of the plate must be 4 in. by 8 in. The plates will be made from a material

with a density of 0.3 lb./in.^{3}. How thick must each plate be?

5. A manufacturer of 1-ft. diameter disc weights receives its metal in 1" x 36" x 36" sheets. The discs are then cut out of the sheets as shown in Figure

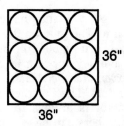

36"

36"

Figure B

B. After the discs are cut, a 1-in. diameter hole is drilled in the center of each.
a.) What is the total volume of steel wasted in the process for each sheet?
b.) If the steel has a specific weight of 0.3 lb./in.^{3}, what is the total weight of the steel wasted with each sheet?

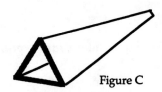

Figure C

6. A high jump bar is usually a hollow triangular cylinder made out of a lightweight metal like aluminum. One particular bar has a base width of 4 cm (0.04m) and a height of 3.5 cm. The base width of the hole is 3.6 cm and its height is 3.1 cm.
a.) If the bar is 400 cm (4m) long, what is the volume of aluminum necessary to manufacture the bar?
b.) If the density of aluminum is 2.71 g/cm^3, what is the mass of the bar?

7. The equation for the volume of a sphere is $V = \frac{4}{3}\pi r^3$. What is the radius of a 10-lb. shot put? ($\gamma = 0.3$ lb./in.^{3})

8. A manufacturer has been making discs with a weight W, radius r_1, a thickness t_1, and a specific weight of γ. It decides it

and a specific weight of γ. It decides it would be easier to make the weights with a new thickness of sheet metal, t_2. What should the new radius, r_2, be in terms of r_1, t_1, and t_2 to maintain the same weight?

3.5 EXPONENTIAL NOTATION

As you may have noticed, the numbers used in material mechanics can get rather large. The load on the column of a building can easily be 100,000,000 lb. or the stress on a part can be 50,000 psi. To make it easier to work with large numbers like this, scientists have developed a format to express large numbers. The format is known as *scientific* or *exponential notation*. In exponential notation, numbers are expressed as the product of two factors. The first factor is always a number between 1 and 10 (or -1 and -10). The second factor is always the number 10 with an exponent. Both very large and very small numbers can be expressed with exponential notation. Below are examples of how very large and very small numbers are expressed with exponential notation.

$$5,638,000 = 5.638 \times 10^6$$

$$0.0004927 = 4.927 \times 10^{-4}$$

In exponential notation the "$\times$" symbol is always used to indicate multiplication. Dots to indicate multiplication are avoided in exponential notation to prevent confusion with the decimal points.

Exercise Set 3.5

Convert the following numbers into exponential notation:

1. 0.0000246

2. -567,100

3. 941,000,000

4. 0.294

5. -94,221

6. 26.7

7. 0.0000004

8. 0.0347

Convert the following numbers into the standard form:

9. 2.09×10^5

10. -9.844×10^{-9}

11. 5.243×10^2

12. 7.543×10^6

13. -3.45×10^{-4}

14. 2.019×10^3

15. 9.008×10^1

16. 5.67×10^{-8}

Perform the following calculations and express the answer in terms of exponential notation:

17. $0.0250 \times -824,100$

18. $-70,400 \div 0.0000176$

19. $-6.75 \times 10^{-9} \div 3.75 \times 10^{-8}$ 21. $0.294 \div 4.9 \times 10^2$

20. $9.08 \times 10^1 \times 2.09 \times 10^5$ 22. $1.25 \times 10^3 \times 940{,}000{,}000$

REFERENCES:

F. Beer and E. Johnston, Jr., *Mechanics of Materials*, New York, McGraw-Hill, 1981. All of the references and formulas relating to the field of material mechanics.

M. Keedy, *Intermediate Algebra*, Reading, Massachusetts, 1980. Formulas for exponents.

Chapter 4

STATISTICS

Statistics are math tools that help us understand the meaning of collections of numbers. By analyzing the numbers (usually called *data*), we can draw conclusions on what is being studied. For example, a basketball team's stat keeper will record points, rebounds, assists, and other statistics so that the coaches can decide which players are doing the best.

When statistics are used in science or business, a survey is usually conducted on a small portion of the total number of items to make estimates on the total number of items. In statistics, the portion which is studied is known as the *sample*. The total number of items is the *population*. For example during an election, a small portion of all of the potential voters are asked on how they plan to vote. This is done to predict who's going to win. In this example, the people who are polled are the sample. All of the people who finally vote make up the population.

Statistics are used somewhat differently in sports than they are in most other areas. In sports, data is collected during the entire competition and for every event. Statistics in sports, therefore, usually study the entire population rather than just a sample. Since the entire population is studied, statistics in sports are usually used to estimate performance in the future. For example a college scout may scan the statistics of high school teams to find the best prospective players. Although there are some differences between sports statistics and statistics used in most other areas, there are enough similarities to make them worth studying.

4.1 STATISTICAL MEASURES

To begin studying statistics, we need to learn some of the terms. Many of the standard statistics terms are different from the ones used in sports.

4.1.1 Dichotomous Populations

Like other types of statisticians, sports statisticians often have an interest in following events that have only two possible outcomes. In statistics, the two outcomes are called *successes* or *failures*. When each element of a population can have only two outcomes, it is called a *dichotomous* population. For example, a basketball player either makes or misses (succeeds or fails at) a free throw. The number of made free throws divided by the total number of attempts is the statistic known as the free throw percentage. Table 4.1 shows other types of dichotomous populations in sports.

Statistic	Population	Success	Failure
Percent win	Games played	win	loss
Batting average	At-bats (excluding walks)	base hit	out
Field goal percentage (basketball)	Field goal attempts	made basket	miss

Table 4.1 Dichotomous Populations

4.1.2 Numerical Populations

While the elements that make up a dichotomous population can be only one of two things, the elements that are used to determine the average number of points per game make up a different type of population. The number of points scored in a game can be almost any number, 12, 15, 43, 6, and so on. When the elements of a population can have a number of different values, the population is said to be a *numerical* population. Table 4.2 shows some examples of statistics with numerical populations:

Statistic	Population
Yards per carry	Rushing gain or loss (in yards) on each carry
Earned run average	Number of earned runs given up in each game pitched
Average punting distance	Distance football travels (in yards) in each punt

Table 4.2 Numerical Populations

4.1.3 The Mean

As in other areas, the most commonly used statistic in sports is the average. In statistics, the average of a set of numbers is known as the *mean*. The equations used to calculate specific types of averages are shown below:

$$\text{Battting average} = \frac{\text{Number of base hits}}{\text{Number of at - bats}} \quad \text{(walks excluded)}$$

$$\text{Points per game} = \frac{\text{Total points scored in games}}{\text{Number of games}}$$

Mathematicians use very precise terms to define the mean of a sample or population. The mean of a set of numbers is often represented by putting a bar over the variable. The equation for the mean shows some new tools which are often used in mathematics.

Definition: The mean $\bar{x}$ of set of n numbers, $x_1, x_2, \ldots, x_n$ is given by the equation:

$$\bar{x} = \frac{x_1 + x_2 + \cdots + x_n}{n} = \frac{\sum\limits_{i=1}^{n} x_i}{n} \qquad (4\text{-}1)$$

The "Σ" in the equation is the upper case sigma, the Greek letter for "s." In math, the Σ indicates summation, the addition of a set of numbers. Note that the set of numbers starts at x_1 and ends at x_n. What the "$\sum\limits_{i=1}^{n} x_i$" term is saying is: "Start at x_1 and add all the numbers through x_n." Equation 4-1 says that to get the mean of a set of numbers, add all of the numbers together and divide by n, the number of numbers.

Example: In a softball tournament, Mary bats 10 times with the following results in sequence:

out, base hit, out, out, walk, base hit, base hit, out, out, out

What is Mary's tournament batting average?

Solution: Batting averages are calculated by dividing the number of hits by the number of at-bats, excluding walks. Since batting averages don't include walks, her walk is ignored. The actual set of elements used to calculate the batting average is therefore 9 at-bats. To help understand the definition of the mean, we'll use the formal equation (Eq. 4-1) to solve the problem. The number set is dichotomous and can be tabulated as follows where 1 equals a hit and 0 equals an out.

x_1	x_2	x_3	x_4	x_5	x_6	x_7	x_8	x_9
0	1	0	0	1	1	0	0	0

$$\bar{x} = \frac{\sum\limits_{i=1}^{9} x_i}{9} = \frac{x_1 + x_2 + \cdots + x_9}{9}$$

$$= \frac{0+1+0+0+1+1+0+0+0}{9} = \frac{3}{9} = \frac{1}{3}$$

$$= 0.333\ldots$$

WITHDRAWN

Mary's tournament batting average is .333.

Example 4.2: In 5 warm up throws before a meet, Reggie throws his shot put the following distances in feet:

$$51.3, 63.6, 54.1, 60.3, 58.2$$

What is Reggie's average throw distance in the five throws?

<u>Solution:</u> The throwing distances make up a numerical population and can be tabulated as follows:

x_1	x_2	x_3	x_4	x_5
51.3	63.6	54.1	60.3	58.2

$$\bar{x} = \frac{\sum_{i=1}^{5} x_i}{5} = \frac{x_1 + x_2 + \cdots + x_5}{5}$$

$$= \frac{51.3 + 63.6 + 54.1 + 60.3 + 58.2}{5} = \frac{287.5}{5}$$

$$= 57.5$$

Of the five shot put throws, Reggie's average distance is 57.5 feet.

4.1.4 The Median

In addition to the mean, there are other types of statistics that can help you understand a set of data. The following example shows how you might need a different type of measure.

Example 4.3: A punt returner would like to know how deep he should play in the upcoming game. Specifically, he wants to know how far he can expect the punter to kick in the game. He asks a trainer to watch the videotape of the opponent's last game and record the punting distances to determine the average punt distance during the game. The resulting data, yards, are as follows:

$$46, 52, 48, 8, 50, 54, 50$$

<u>Solution:</u> The mean of the data is:

$$\bar{x} = \frac{46 + 52 + 48 + 8 + 50 + 54 + 50}{7} = \frac{308}{7}$$

$$= 44$$

The average punting distance 44 yards.

From looking at the data however, the trainer should realize that the punter kicked only one punt less than 46 yards. If the punt returner positions himself at 44 yards from the punter, he will be playing too shallow. The fluke of the 8-yard punt throws off the data to where the mean is misleading. To better position himself, another type of measure should be studied.

Since samples and populations can often include values which are extreme enough to make the mean worthless, the middle value of the population is often a better measure. In statistics, the middle value is known as the *median*. The median indicates the value of which there are as many elements less than it as there are greater than it.

Definition: Given the set $x_1, x_2 \ldots, x_n$ where the observations are arranged in increasing order (from most negative to most positive) the median is given by

When n is odd:

    x = single middle value in the ordered list

When n is even:

    x = the average of the two middle values in the ordered list

---

If you arrange the punts in ascending order,

$$8, 46, 48, 50, 50, 52, 54$$

The median value of 50 is obtained by inspection.

4.1.5 Maximum, Minimum, and Range

Other important statistics include the *maximum* and *minimum*. The maximum is the greatest value in the set and the minimum is the least value in the set. The difference between the maximum and the minimum is known as the *range* of the set.

In sports which are scored by a panel of judges, such as gymnastics, diving, and figure skating, the highest and lowest scores are often thrown out. This is done in the Olympics to minimize any political biases that may exist in the judging.

Example 4.4: In the balance beam competition, Lisa receives the following scores from the panel of judges:

Japan	Denmark	Australia	Argentina
7.8	6.9	7.4	8.1

Determine the maximum and minimum scores. If the maximum and minimum scores are discarded, what is Lisa's score in this event? What is the range of the scores?

Solution: The maximum and minimum values can be determined by inspection. The maximum score is from the Argentinean judge and is 8.1. The minimum score is from the Danish judge and is 6.9.

If the maximum and minimum scores are discarded, the average score is:

$$\text{average score} = \frac{\sum \text{scores}}{\text{number of scores}}$$

$$= \frac{7.8 + 7.4}{2}$$

$$= 7.6$$

Lisa's average score is 7.6.

The range is obtained by subtracting the minimum from the maximum.

$$\text{range} = 8.1 - 6.9$$

$$= 1.2$$

The range of Lisa's scores is 1.2.

Exercise Set 4.1

1. As a running back, the yardage gains in Earl's carries are:

4, 16, -2, 24, 6, 1, 4, 42, 12, 2, 16, 7, 3, 9, 5, 6.

a.) What was his total rushing for the game in yards?
b.) What was his average yards per carry?

2. Danielle's scoring in basketball so far this season has been:

16, 12, 22, 15, 8, 15, 26, 21, 23

a.) Is this population dichotomous or numerical?
b.) What is her average points per game?
c.) What is her median score?
d.) What is the range of scores?

3. During a softball tournament, Nancy bats as follows:
out, out, hit, walk, hit, out, walk, out, hit, hit, out, out
What was her tournament batting average?

4. Given the win (W) and loss (L) data in the table, complete the percent win (Pct.) column:

Team	W	L	Pct.
Houston	22	1	
Utah	17	8	
San Antonio	15	10	
Denver	10	13	
Minnesota	8	16	
Dallas	1	23	

Does the data represent a dichotomous or numerical population?

5. A left fielder wants to know how deep to play a batter who has been hitting exceptionally well. From the video tapes of the batter's games, the distances of his hits to left field are recorded. The data are as follows with the distances in feet:

203, 198, 212, 208, 192, 121, 193, 207, 230

a.) What is the player's mean distance hit?
b.) What is the median length hit?

6. A group of forest rangers conduct a study of trees which survive a forest fire. Borings are made in the trees to enable the rangers to count the rings of the trees to determine their ages. Of the trees which are sampled, the following age data are obtained:

57, 163, 94, 201, 211, 86, 127, 144, 88, 181, 276, 172, 219, 154, 43, 87, 123, 132, 198, 67

a.) What is he sample mean?
b.) What is the median age?
c.) What is the minimum, maximum, and range of the sample?

7. Because of budget cuts, a transit agency is considering using smaller, more economical buses on some of its routes. To determine if it can use a smaller bus on one route, it counts the number of passengers along the busiest section of the route during rush hour. The passengers are counted every day for a week. The resulting data are as follows:

44, 32, 38, 40, 35, 12, 4

Determine the mean, median, maximum, minimum, and range of the sample.

Find the mean, median, maximum, minimum and range for the following populations:

8. 7, 3, 21, 6, 13, 5, 2, 7

9. 106, 27, 304, 211

10. 41, 23, 6, 85, 47, 38, 19

11. 2, 18, 1, 6, 34, 29

12. 5839, 2411, 2705, 6094, 3128, 1226

4.2 ELECTRONIC SPREADSHEETS

People who work with large amounts of statistics, in sports or other areas, need to be able to do mathematical calculations on large amounts of data. The easiest way to perform large amounts of calculations is with computers. Although there are computer programs written especially for statistics, some fields use highly specialized calculations for which there are no programs. People in specialized fields may have to do their own programming to get the statistics they need. Perhaps the easiest way to program your own calculations is with *electronic spreadsheets*. Electronic spreadsheets are computer programs that are very easy to use and relatively easy to learn how to use. The tables you read in the sports page everyday are probably made with spreadsheets. There are several spreadsheet programs available written by different companies. Each program is a little different, but all operate on basically the same principles.

Spreadsheets divide the computer screen into a grid of rows and columns. The rows and columns in Figure 4.1 show what you would see when the program is running. The rows and columns of the spreadsheets give us an orderly way to view and work with our data. Each box created by the rows and columns is known as a *cell*. The spreadsheet's cells may contain numbers, text, or mathematical formulas. You can move between the cells by using the computer's arrow keys or with a mouse if one is available.

G3											
	A	B	C	D	E	F	G	H	I	J	K
1											
2											
3											
4											

Figure 4.1 Blank Spreadsheet

Most spreadsheet programs label the columns with letters and the rows with numbers. The different cells are identified by column and row designations. The cell highlighted in Figure 4.1 has the designation G3. When a cell is highlighted on the screen, there is usually an area at the top of the spreadsheet which shows the contents of the cell. Across the top of the screen, you'll see a *menu* (not shown here). The menu is a list of commands which run the program.

The following discussion shows how to set up a table which calculates some batting statistics of a baseball team. The purpose of this discussion is to give an introduction to spreadsheets. If you would like to actually follow along on a computer, you will need to refer to the program's manual to get started or ask someone familiar with the program. In this table, we're going to display the number of hits, walks, and at-bats from a game next to the players' names. We'll then create a column which calculates their batting averages. At the bottom, we'll put a row of team statistics.

We'll start by adding the text of the table. We'll place the players' names in a column on the left side and place the titles of the statistics across the top row. Unless told to do otherwise, the computer will align text entries on the left side of the cell. You may need to adjust the width of the column to fit the names. Most spreadsheets let you do this by simply grabbing the edge of the top of the column with the mouse and moving it. In other programs, you'll need to use the commands from the menu.

B4	Lee					
	A	B	C	E	F	G
1		Player	Hits	Walks	At-bats	Batting Ave.
2		Johnson				
3		Allen				
4		Lee				
5		Malletti				
6		Marshall				
7		Robinson				
8		Ng				
9		Reynolds				
10		Calavera				
11		Team Totals				

Figure 4.2 Adding Text

Next we add the numbers from the game. Unless told to do otherwise, the computer will align the numbers on the right side of the cell.

E7	3				
	A	B	C	E	F	G
1		Player	Hits	Walks	At-bats	Batting Ave.
2		Johnson	3	1	7	
3		Allen	2	2	7	
4		Lee	2	0	7	
5		Malletti	3	1	6	
6		Marshall	0	2	6	
7		Robinson	2	3	6	
8		Ng	3	1	6	
9		Reynolds	2	0	6	
10		Calavera	2	1	6	
11		Team Totals				

Figure 4.3 Adding Numbers

Now that the text and numbers are entered, we can enter the formulas that will calculate the batting averages and team statistics. The batting averages will go in column G. The formula to calculate batting averages is:

$$\text{batting average} = \frac{\text{number of base hits}}{(\text{total number of at - bats}) - \text{walks}}$$

To get this formula into the cells of column G, we substitute the appropriate cell locations for the variables in the equation. Column C has the hits; column F has the at-bats; and column E has the walks. Therefore, in cell G2 we would enter:

$$=C2/(F2-E2)$$

Some programs may use a plus sign, "+", or a "@" symbol in the formula instead of the equal sign, "=". When the formula has been entered correctly, you will see it quickly displays the batting average.

G2	=C2/(F2-E2)				
	A	B	C	E	F	G
1		Player	Hits	Walks	At-bats	Batting Ave.
2		Johnson	3	1	7	0.50000
3		Allen	2	2	7	
4		Lee	2	0	7	

Figure 4.4 Adding Formulas

You may need to adjust the format of the cell so that it displays the standard number of decimal places for a batting average (3). The decimal format command can be found in the menu.

G2	=C2/(F2-E2)				
	A	B	C	E	F	G
1		Player	Hits	Walks	At-bats	Batting Ave.
2		Johnson	3	1	7	0.500
3		Allen	2	2	7	
4		Lee	2	0	7	

Figure 4.5 Adjusting Decimal Places

Now that you have the formula for cell G2, it can be easily copied to the rest of the cells in column G. Some spreadsheets do this with a Copy command from the menu; others can complete the column with a Fill Down command. Unless told to do otherwise, the computer will know that when this formula is copied to other rows, it needs to use the data from the same row and not the data from the row it was copied from. If you look at the top of the spreadsheet as you arrow down column G, the column letters in the formula will stay the same but the row numbers will change. Compare the cell box at the top of the screen in Figure 4.6 with the cell box of Figure 4.5.

G3	=C3/(F3-E3)				
	A	B	C	E	F	G
1		Player	Hits	Walks	At-bats	Batting Ave.
2		Johnson	3	1	7	0.500
3		Allen	2	2	7	0.400
4		Lee	2	0	7	0.286
5		Malletti	3	1	6	0.600
6		Marshall	0	2	6	0.000
7		Robinson	2	3	6	0.667
8		Ng	3	1	6	0.600
9		Reynolds	2	0	6	0.333
10		Calavera	2	1	6	0.400
11		Team Totals				

Figure 4.6 Copying the Formula

The final step is to add the summary row at the bottom of the table. Spreadsheet programs usually come with some pre-programmed functions. One typical function is the SUM() function. If we want to total the number of hits of the team, we can go to cell C11 and enter:

=SUM(C2..C10)

This will calculate the sum of the contents of rows 2 through 10 in column C. In most spreadsheets, you only need to enter the SUM() and then point to the range of cells with the arrow keys or mouse. The computer will know that this is the range of the cells to sum. The formula can then be copied to the at-bat and walk columns.

C11	=SUM(C2..C10)				
	A	B	C	E	F	G
1		Player	Hits	Walks	At-bats	Batting Ave.
2		Johnson	3	1	7	0.500
3		Allen	2	2	7	0.400
4		Lee	2	0	7	0.286
5		Malletti	3	1	6	0.600
6		Marshall	0	2	6	0.000
7		Robinson	2	3	6	0.667
8		Ng	3	1	6	0.600
9		Reynolds	2	0	6	0.333
10		Calavera	2	1	6	0.400
11		Team Totals	19	11	57	

Figure 4.7 Calculating Team Totals

The team batting average can be calculated in several ways. One method is to copy the batting average formula into cell G11. A second method is to use another pre-programmed function. The Average() function is available on most spreadsheets. You could enter = Average (G2..G10) into cell G11 to get the average batting average of the team.

G11	= Average (G2..G10)				
	A	B	C	E	F	G
1		Player	Hits	Walks	At-bats	Batting Ave.
2		Johnson	3	1	7	0.500
3		Allen	2	2	7	0.400
4		Lee	2	0	7	0.286
5		Malletti	3	1	6	0.600
6		Marshall	0	2	6	0.000
7		Robinson	2	3	6	0.667
8		Ng	3	1	6	0.600
9		Reynolds	2	0	6	0.333
10		Calavera	2	1	6	0.400
11		Team Totals	19	11	57	0.413

Figure 4.8 Completed Spreadsheet

Once the spreadsheet is complete, you'll notice that if you change any of the numbers, the computer will instantly recalculate the statistics. This is one of the greatest benefits of spreadsheets. Once completed, spreadsheet files can be saved on a disk. If you want to calculate the same averages for a different game or a different team, all you need to do is re-enter the different names and numbers. The formulas will remain and the computer will automatically calculate the new statistics.

As you'll see in several areas of this book, many types of problems are best solved by putting the data and calculations into tables. The orderliness, speed, and flexibility offered by spreadsheets make solving these types of problems much easier.

As stated earlier, the purpose of studying algebra is largely to be able to manipulate mathematical formulas. By understanding how computers can quickly execute these formulas, you can better appreciate the power of mathematics.

Exercise Set 4.2

The following tables contain raw data to be calculated with an electronic spreadsheet. Use a spreadsheet program to calculate the missing statistics.

1. Baseball tournament batting

Game 1:

Player	Hits	Walks	At-bats	Batting Ave.
Crane	2	2	6	
Franco	1	0	6	
Lamb	2	1	6	
Carr	0	2	6	
Owens	3	1	5	
Clifford	1	0	5	
Griffith	4	1	5	
Mendez	2	0	5	
Davis	0	0	5	
Team Totals				

Game 2:

Player	Hits	Walks	At-bats	Batting Ave.
Crane	3	1	6	
Franco	2	1	6	
Lamb	4	0	6	
Carr	1	3	6	
Owens	0	2	6	
Clifford	2	1	5	
Griffith	3	2	5	
Mendez	3	1	5	
Davis	4	0	5	
Team Totals				

Game 3:

Player	Hits	Walks	At-bats	Batting Ave.
Crane	4	2	6	
Franco	3	1	6	
Lamb	3	2	6	
Carr	2	0	6	
Owens	2	2	6	
Clifford	3	3	6	
Griffith	1	2	5	
Mendez	3	1	5	
Davis	2	2	5	
Team Totals				

Game 4:

Player	Hits	Walks	At-bats	Batting Ave.
Crane	1	3	6	
Franco	0	4	6	
Lamb	1	3	6	
Carr	4	0	5	
Owens	1	2	5	
Clifford	1	3	5	
Griffith	2	2	5	
Mendez	3	1	5	
Davis	2	2	5	
Team Totals				

2. Uneven Parallel Bar Competition (all scores counted)

Gymnast	Score by Judge				Average Score
	Kenya	Philippines	Costa Rica	Austria	
Garcia	7.4	7.1	6.8	8.0	
Taguchi	7.8	8.2	7.4	8.0	
Peng	6.4	7.2	6.6	6.3	
Ahmed	7.5	7.1	6.9	7.4	
Bryzek	8.1	8.4	7.9	8.8	
Torres	7.8	7.3	7.4	6.9	
Shvetsoff	8.2	7.5	8.6	8.0	
Narain	6.9	8.1	7.5	8.2	

3. Hockey Team Standings

Team	Wins	Losses	Ties	Goals	Pct. Win	Goals/Game
Pittsburgh	21	12	10	52		
Montreal	21	16	7	49		
Boston	19	15	8	46		
Buffalo	10	19	4	44		
Quebec	17	22	5	39		
Hartford	17	24	3	37		
Ottawa	8	34	5	21		

4. Football Passing Statistics

Quarterback	Attempts	Completed	Total Yards	Pct. Complete	Yd./Attempt
Robinson	345	211	2467		
Taylor	413	246	2961		
Smith	213	123	1702		
Lopez	334	191	2378		
Richards	286	166	2125		
Jordon	279	159	1905		

5. Basketball Shooting Statistics   M - Made   A - Attempts   % - Percent Made

Player	3-Point			2-Point			Free Throw			Total Points
	M	A	%	M	A	%	M	A	%	
Unruh	7	14		9	17		5	6		
Rollins	2	2		8	20		7	14		
Farris	1	3		4	9		2	2		
Driscol	4	5		6	11		0	0		
Lang	5	6		12	16		6	7		
Keefe	1	4		8	18		4	9		
Edwards	2	3		2	3		2	3		
Graham	4	4		4	8		6	6		
Worthy	0	0		7	11		4	8		
Campbell	0	2		8	15		2	2		

6. A manufacturer of weight lifting equipment wants to produce a line of one-arm barbells for every 5-lb. increment between 20 and 100 lb. The weights will consist of two discs attached to a 3-lb handle. The weights of 20 to 50 lb. will have a diameter of 6 in. The weights of 55 to 100 lb. will have a diameter of 8 in. The discs will be made from steel with a specific gravity of 0.3 lb/in.^{3}. Each weight will have a different thickness. Develop a spreadsheet which determines the required thickness of the discs of each weight.

Chapter 5

THE CARTESIAN COORDINATE SYSTEM

In Chapter 2, we used a number line to keep track of the progress of the football down the field. The number line gives us a sideline view of the field. Unfortunately, just as the sideline gives you a very limited view of the game, number lines are very limited in how much they can help us in math. This chapter shows how math analyzes relationships in two or three dimensions. Again, the football field presents a good comparison with the system used in mathematics. But now we need to rotate the field to give us a top view.

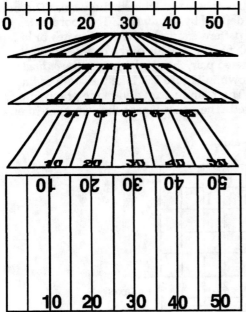

The two-dimensional graphing system used in math is known as the *Cartesian Coordinate system*. The system was developed by a French mathematician/philosopher named René Descartes.

Examining a pass pattern gives us a good start at understanding the coordinate system. On the number line, a ten-yard completed pass (starting at 0) is shown like this:

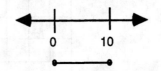

Fig. 5.1 One-Dimensional Pass Pattern

Again, the number line gives us a sideline view of the play. The one-dimensional graph can't show what kind of pass pattern the receiver ran.

On the Cartesian coordinate system, a down and out pattern (10 yards down and 5 yards out) can be shown like this:

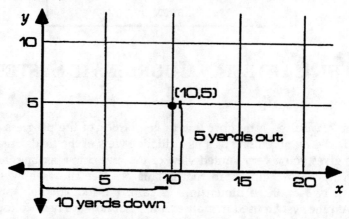

Figure 5.2 Two-Dimensional Pass Pattern

In the graph, you can see that another number line has been added perpendicular to the original number line to measure distance in the other direction. The horizontal line is usually labeled the *x-axis* and the vertical axis is usually labeled the *y-axis*. The point where the down and out pattern ends can be defined by *coordinates*. The coordinates give the *x*- and *y*-values of the point. In the pass pattern example, the coordinates or *ordered pair*, of the point where the pattern ends is (10, 5). The *x*-value of the ordered pair, the 10, is known as the *abscissa*. The *y*-value of the ordered pair, the 5, is known as the *ordinate*.

The full Cartesian coordinate graph is shown in Figure 5.3. Like the number line, the *y*-axis shows positive and negative numbers. The *x*- and *y*-axes divide the graph into four quadrants. Each of the quadrants is numbered. The point where the two axes intersect (the zero point) is called the *origin*.

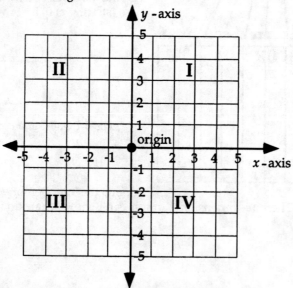

Figure 5.3 The Cartesian Graph

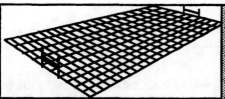

Comparisons between the football field and the Cartesian Coordinate system are actually not too farfetched. At one time, the football field looked very much like the Cartesian graph. Near the turn of the century, as football was evolving from a soccer-like game, it became legal to pass the ball forward. This was considered to be such a substantial change that efforts were made to limit the passing game. A rule was made to limit the movement a quarterback could only make to five yards to the left or right before passing the ball. To help referees enforce the 5-yard limit, the field was striped every five yards in both directions. The striping gave the field the appearance of a "*gridiron*", a name which is still sometimes used to describe the field. The 5-yard limit was later eliminated and the extra lines were removed.

5.1 PLOTTING POINTS

A common use of graphs in the business world is to plot trends. Sports statisticians also plot trends sometimes. The following example shows how a graph can be used to help predict performances in track and field events.

<u>Example 5.1:</u> The data in Table 5.1 shows the progress of world record in the men's high jump. Plot the heights by year and determine a rough approximation of what the record will be in the year 2000.

Year	1912	1924	1934	1941	1953	1957	1960	1963	1971	1976	1980	1985	1989
Heig ht (m)	2.00	2.03	2.06	2.11	2.12	2.16	2.22	2.28	2.29	2.32	2.36	2.41	2.44

Table 5.1 World Records in Men's High Jump

<u>Solution:</u> The data can be plotted as follows:

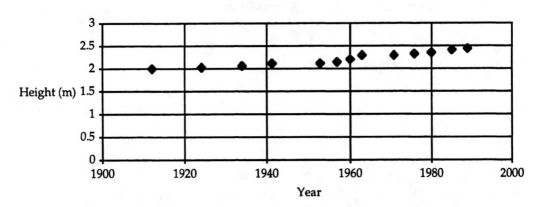

Figure 5.4 Plot of Men's High Jump Records

If we connect the dots with a line, we can guess that the record for the high jump will be slightly over 2.5 meters at the year 2000.

In the down and out pattern example, the x- and y-axes both indicate yards. In the high jump record example, the x-axis indicates years and the y-axis indicates heights. The units of the axes of Cartesian graphs will often be different. We say that an axis is *graduated* when it has markings to show the values. For some problems, it makes sense to number the markings on the axes differently. In the high jump example, the y-axis is graduated from 0 to 2.5 meters while the x-axis is graduated from 1910 to 2000. Graduating the axes differently is sometimes done to better present the data. If the axes on the high jump problem were graduated the same, the y-axis would have a height of 2.5 units and the x-axis would have a width of 90 units (2000-1910 = 90). As a result, the graph would be extremely short and wide and would be hard to fit on the page. To make the graph more readable, the length of one meter in the y-direction is about the same as the length of 10 years in the x-direction.

5.1.1 Surveying

Another common use of coordinate systems is to define physical locations. Surveyors and engineers use coordinate systems to keep track of existing features in an area and to position new structures and facilities. In surveying graphs, the y-axis usually represents the north-south direction with north in the positive direction and south in the negative direction. The x-axis represents the east-west direction with west in the negative direction (to the left) and east in the positive direction (to the right). When surveying coordinates are tabulated, the letters "N" and "E" are added to differentiate the north from the east coordinates.

Example 5.2: An aerial photograph of an area is taken. Surveyors later overlay the photograph with the appropriate coordinate system and create the figure below. Determine a coordinates of the following features:

    The intersection of Jameson Rd. with First Street
    The intersection of Jameson Rd. with Second Street
    The fork in Turner River
    The Jameson Road bridge over Turner River
    The Post Office
    The east end of the shopping center

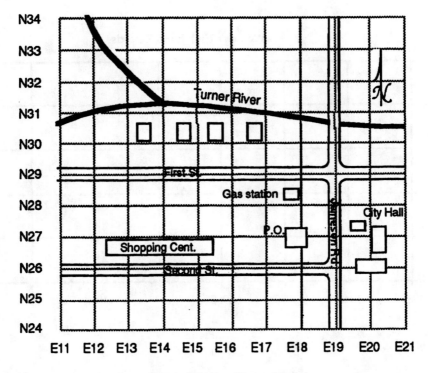

Figure 5.5 Coordinate Map

Solution: The coordinates are determined by inspection as follows:

Location	Coordinates[1]
The intersection of Jameson Road with First Street	(E19, N29)
The intersection of Jameson Road with Second Street	(E19, N26)
The fork in Turner River	(E13.9, N31.3)
The Jameson Road bridge over Turner River	(E19, N30.7)
The Post Office	(E18, N27)
The east end of the shopping center	(E15.4, N26.7)

Exercise Set 5.1 - Plotting Points

1. The data in the table below represents the world's record in the women's high jump.

Year	1932	1939	1943	1951	1957	1961	1971	1976	1983	1986	1987
Height	1.65	1.66	1.71	1.72	1.77	1.91	1.92	1.96	2.04	2.08	2.12

Plot the record heights versus the years in a graph as done in Example 5.1. Graduate the axes of the graph as shown in Figure A.

---

[1] It should be noted that engineers and surveyors traditionally place the north-south values of a coordinate first and the east-west value second. To illustrate a practical use of coordinates in the standard xy sequence, the east-west values have been placed first.

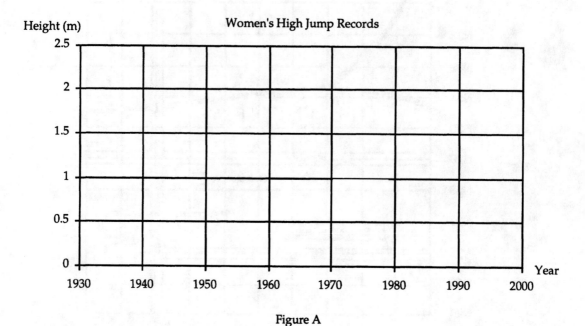

Figure A

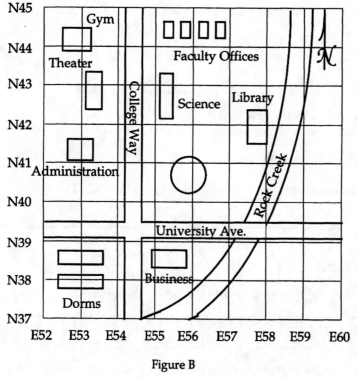

Figure B

2. The map in Figure B shows the layout of a college campus. Determine the coordinates of the following locations:

a.) Library

b.) Business building

c.) bridge over Rock Creek

d.) southernmost dorm

e.) Administration building

f.) westernmost faculty office building

g.) Gym

h.) intersection of University Ave. and College Way

Problems 3-12: Determine the coordinates of the points in the graph in Figure C. Also determine in which quadrant, if any, the point is in.

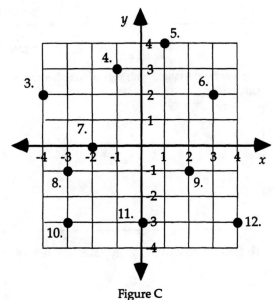

Figure C

Plot the following points on a graph:

13. (3,7)
14. (6,-1)
15. (0,-7)
16. (5,-4)
17. (-9,1)
18. (-12,-4)
19. (3,0)
20. (0,0)

5.2 DISTANCE

When coordinate systems are used to map an area, it is sometimes necessary to determine the distance between two points. There is a formula to calculate the distance between any two points on a plane when you know their coordinates. The formula is based on a more common equation known as the *Pythagorean Theorem*.

5.2.1 The Pythagorean Theorem

A triangle which has two sides perpendicular to each other is known as a *right* triangle. The two sides which are perpendicular to each other are known as the *legs* of the triangle. The third and longest side of the triangle is known as the *hypotenuse*.

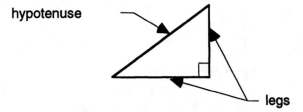

hypotenuse

legs

Figure 5.6 Right Triangle

If we know the lengths of any two sides of a right triangle, we can find the length of the other using the Pythagorean Theorem. The theorem is named after the Greek mathematician, Pythagorus, who discovered the relationship.

Pythagorean Theorem

For any right triangle with legs of length a and b, the length of the hypotenuse, c, is given by the equation:

$$a^2 + b^2 = c^2$$

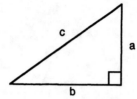

To find the length of the hypotenuse, we solve the equation for c.

$$c^2 = a^2 + b^2$$
$$c = \pm\sqrt{a^2 + b^2}$$

(5-1)

Since lengths are always positive, we don't need to consider the negative value of the square root.

$$c = \sqrt{a^2 + b^2}$$

Example 5.3: While playing right field, Joe throws the ball from a position 180 feet down the foul line and 60 feet from the foul line as is shown in Figure 5.7. How far is the throw to home plate from this position?

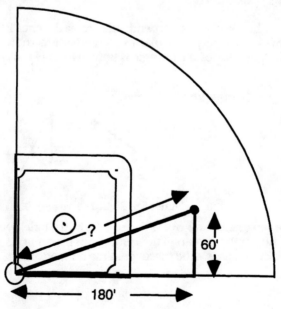

Figure 5.7 Throwing Distance Calculation

Solution: To find the distance, we form a triangle that has legs of length 180 feet and 60 feet. The throw to home will equal the length of the hypotenuse of this triangle. Using the Pythagorean Theorem, the hypotenuse of the triangle is:

$$h = \sqrt{180^2 + 60^2}$$

$$= \sqrt{36,000}$$

$$\approx 190$$

Joe has a throw of about 190 ft. to home plate.

What would be the distance to second base from Joe's position? Finding the distance to second helps show how coordinates can be used to solve these types of distance problems. The Pythagorean Theorem has been worked into a formula which can be used to calculate the distance between any two points on a graph. If we look at the baseball diamond as if it were superimposed on a Cartesian graph with home plate at the origin and the right field foul line as the x-axis, the coordinates of Joe's position would be (180, 60). The coordinates of second base would be (90, 90).

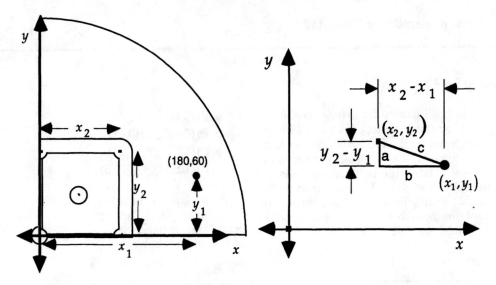

Figure 5.8 Throwing Distance to Second Base

From the figure on the right we can see that a triangle can be formed having legs where $a = y_2 - y_1$, and $b = x_2 - x_1$. The Pythagorean Theorem can then be applied.

$$c^2 = (x_2 - x_1)^2 + (y_2 - y_1)^2$$

If the equation is solved for c, we get the standard equation to calculate the distance, d, between to points on a plane:

$$d = \sqrt{(x_2 - x_1)^2 + (y_2 - y_1)^2} \qquad (5\text{-}2)$$

Using Equation 5-2, the throwing distance, d, is therefore be calculated as follows:

$$
\begin{aligned}
d &= \sqrt{(x_2 - x_1)^2 + (y_2 - y_1)^2} \\
&= \sqrt{(90 - 180)^2 + (90 - 60)^2} \\
&= \sqrt{(-90)^2 + 30^2} \\
&= \sqrt{8,100 + 900} \\
&= \sqrt{9,000} \\
&\approx 94.87
\end{aligned}
$$

The throw to second base is about 95 ft.

Exercise Set 5.2

1. In a softball game, the catcher chases a pop fly near the backstop. If an xy plane is superimposed on the field with home plate at the origin as in Example 5.3, the location where she catches the ball can be defined by the coordinates (-6,-16) with units in ft. From this point, how far is the throw to each of the four bases? (The bases on a softball field are 65 ft. apart.)

4. $P_1(5,1)$; $P_2(12,-23)$
5. $P_1(0,-17)$; $P_2(0,41)$
6. $P_1(12,-1)$; $P_2(4,13)$
7. $P_1(10,-11)$; $P_2(-2,-3)$

8. A group of landscape architects is designing a park. The piping and sprinkler heads of the park's irrigation system will be laid out as shown in Figure B.

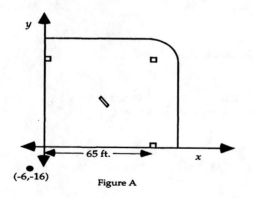

$(-6,-16)$

Figure A

Find the distance between the two points:

2. $P_1(2,7)$; $P_2(6,4)$
3. $P_1(11,-7)$; $P_2(5,1)$

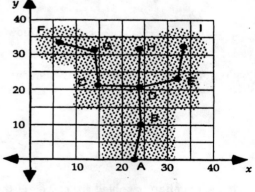

Figure B

The coordinates of the sprinkler heads are:

A (23,0)	E (32,23)	I (33,33)
B (24,10)	F (6,33)	
C (14,22)	G (9,31)	
D (23,21)	H (24,32)	

Determine the amount of linear feet of piping needed for the irrigation system.

5.3 THREE-DIMENSIONAL COORDINATE SYSTEMS

The beginning of this chapter discussed the benefits of watching a football game from a two-dimensional perspective above the field as opposed to a one-dimensional perspective from the sideline. While a two-dimensional perspective is better, it still doesn't give a complete view of the activity on the field. For example, it would be very difficult to tell if field goals were good if you viewed the game from directly above the field. It would be hard to see if the ball passed above or below the crossbar. To keep track of a football game and other activities in the physical world, one really needs to be able to see in three dimensions.

To track relationships with three dimensions, a three-dimensional coordinate system is used. The system adds a new axis, the z-axis, to the xy plane. The x- and y-axes are reoriented somewhat. If the xy plane were considered to be the east/west and north/south directions of a building, the z-axis could be considered to be additional stories on the building. The xyz system is oriented as shown in Figure 5.9.

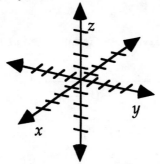

Figure 5.9 Three-Dimensional Axes

Points are plotted in the three-dimensional system with coordinates as they are in the xy plane system. The distance between points is calculated in a similar way to that of the two-dimensional system. The formula for distance between two points (x_1, y_1, z_1) and (x_2, y_2, z_2) is given by the equation:

$$d = \sqrt{(x_2 - x_1)^2 + (y_2 - y_1)^2 + (z_2 - z_1)^2} \qquad (5\text{-}3)$$

<u>Example 5.4:</u> A soccer team is about to do a corner kick. Juan and Matt are two forwards positioned near the goal waiting for the kick. Given their positions and the position of the goalie, Juan's best shot is a kick to the lower left corner and Matt's best shot is a head to the top left corner. As a reference frame, an *xyz* coordinate system is placed on the field with the base of the poles on the *y*-axis and the left pole on the *z*-axis. Table 5.2 and Figure 5.10 show the coordinates of the locations where the shots would begin and the coordinates of their targets. Given the data, determine which player has the shortest shot.

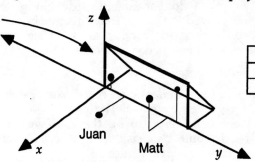

Player	Origin of Shot	Target
Juan	(10,8,0)	(0,2,1)
Matt	(8,20,6)	(0,22,7)

Table 5.2 Player and Target Coordinates

Figure 5.10 Corner Kick

<u>Solution:</u> The distance of each player's shot is calculated using Eq. 5-3.

For Juan:

$$d = \sqrt{(x_2 - x_1)^2 + (y_2 - y_1)^2 + (z_2 - z_1)^2}$$

$$= \sqrt{(0 - 10)^2 + (2 - 8)^2 + (1 - 0)^2}$$

$$= \sqrt{(-10)^2 + (-6)^2 + 1^2}$$

$$= \sqrt{100 + 36 + 1} = \sqrt{137} \approx 11.7$$

For Matt:

$$d = \sqrt{(x_2 - x_1)^2 + (y_2 - y_1)^2 + (z_2 - z_1)^2}$$

$$= \sqrt{(0 - 8)^2 + (22 - 20)^2 + (7 - 6)^2}$$

$$= \sqrt{(-8)^2 + 2^2 + 1^2}$$

$$= \sqrt{64 + 4 + 1} = \sqrt{69} \approx 8.3 < 11.7$$

Matt has the shortest shot.

Exercise Set 5.3

1. When archeologists dig on a site expected to have artifacts, they often create a grid of strings to show coordinates of each artifact they discover. Figure A shows the positions of certain artifacts discovered on an archeological dig. Determine the coordinates of each of items A through I.

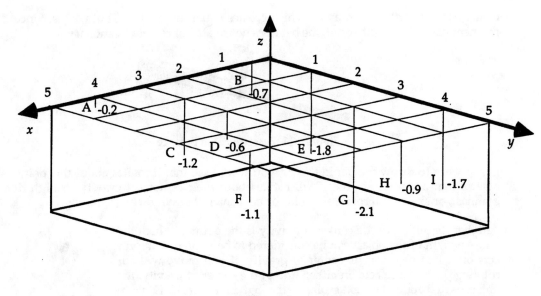

Figure A

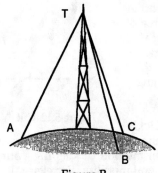

Figure B

2. A radio station wants to build a new transmission tower on a hill. Three guy wires will be used to support the tower as shown in the figure. The guy wires will be attached to the ground at points with coordinates of A(0,-116,11), B(100,58,5), and C(-100,58,8) with units in ft. The coordinates of the point on the tower where the guy wire will attach are T(0,0,212). What is the length of each wire? What is the total length of wire the station will need in linear feet?

Find the distance between the two points:

3. $P_1(0,0,0)$; $P_2(3,3,3)$ 5. $P_1(5,1,27)$; $P_2(12,-23,0)$

4. $P_1(9,0,8)$; $P_2(5,0,-1)$ 6. $P_1(3,-2,7)$; $P_2(0,1,-4)$

5.4 CENTER OF GRAVITY

As shown earlier, one of the main uses of three dimensional coordinate systems is to define points. For scientists, engineers, and athletes, one of the most important points of an object or body is its *center of gravity*. Coordinate systems are often used to define and determine the center of gravity of an object. Although many athletes may not realize it, a familiarity with the concept of center of gravity is important in nearly every sport.

The center of gravity of an object is the point at which the object would balance if it were supported only at this point. For example, if you were to support a baseball bat horizontally with your finger, the point where you finger balances the bat perfectly would be below the center of gravity. The bat balances because the weight on each side of the bat is distributed in away which creates equal moments on both sides. Since the moments cancel each other out, the bat does not rotate but remains stationary.

Figure 5.11 Baseball Bat at Balance

If you were to throw the bat in the air, you would see that it rotates about this point. All objects that are rotating freely will rotate about their centers of gravity. A high diver, gymnast, or skater will rotate about his or her center of gravity when spinning.

Another definition of the center of gravity is the point at which the entire weight of the object can be considered to be located. If a force acts on a free object at its center of gravity, it will move without rotating. If the force acts on either side of the center of gravity, the object will rotate. For example, if the bat in Figure 5.11 were supported at point to the left or right of the center of gravity, the bat would rotate and fall. In a similar way, when a football player makes a sharp cut, he plants one leg and leans in the other direction. This puts his center of gravity on line with the force of his leg. Without doing this, a large moment would be created between the line of force of his leg and his center of gravity and he would lose balance. Similarly, cyclists need to lean to get their centers of gravity on line with the forces that occur when turning a corner.

Figure 5.12 Cut

Athletes that need to be able to turn quickly and keep their balance, know to keep their center of gravity low. Crouching with bent knees is a common method of keeping your center of gravity low and maximizing lateral mobility. For example, a goalie in hockey or soccer who is anticipating a shot will crouch down to be able to move quickly in either direction. A baseball infielder fielding a grounder will do the same thing in case the ball takes a quick bounce. A base runner stays low so that he can either steal or dive back to the base if the pitcher tries to throw him out. A basketball player covering a ball handler will crouch to defend against drives. Similarly, a defensive lineman will crouch to be able to respond to lateral movements of the ball carrier. In all of these examples, athletes are trying to keep their centers of gravity low. When their legs and feet move their bodies, moments are created about their centers of gravity. If they stand erect, the moment arm between their feet and center of gravity will be very large and they may lose their balance. When they crouch down, their center of gravity is lowered and the moment arm between their feet and center of gravity is reduced. This is like when a batter "chokes up" on a bat. When a batter chokes up, the distance between the point where they grip the bat and its center of gravity is reduced. When the moment arm between the grip and the center of gravity is reduced, the bat is easier to swing.

As you might imagine, the center of gravity is always a critical consideration when designing machinery. For example, race cars and sports cars are designed short to have a low center of gravity which allows then to take turns at high speeds.

5.4.1 Determination of the Center of Gravity

The center of gravity of simply shaped objects like rectangular blocks, cylinders, and spheres is very easy to calculate. If they are made of a uniform material, the center of gravity is simply the center of the object. The center of gravity of triangular plate is located a third of the height from its base. The center of gravity of a cone is located a quarter of the height from its base.

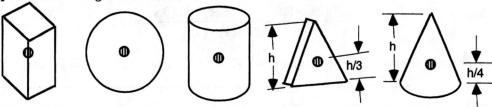

Figure 5.13 Centers of Gravity

The determination of the center of gravity of a complicated object is done by breaking it down into sections and then using a system of coordinates to find the "average" center of gravity of the object. The formula used to determine the y-position of the center of gravity of a complex object is as follows:

$$\bar{y} = \frac{\sum_{i=1}^{n} w_i y_i}{\sum_{i=1}^{n} w_i}$$

where: w = the weight of the element,
y = the height of the element's center of gravity   (5-4)
n = the number of elements

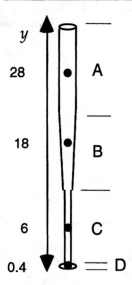

Fig. 5.14 Bat Sections

Example 5.5: The bat in Figure 5.14 is solid wood. To calculate its center of gravity, it is divided into four sections, A, B, C, and D. The axis of the bat is considered to be a y-axis. Sections A and C are both cylinders. Their centers of gravity are therefore at their centers. The cap, Section D, is another symmetrical shape whose center of gravity is at its center. Section B is thicker on one end. This shape is known as a *conical frustum*. (Its center of gravity is determined with a complex formula.) The sections of the bat have the center of gravity data shown in Table 5.3 with the weight, w, in ounces and the y values in inches. Given the data, determine the location of the center of gravity of the bat.

Element	w	y
A	20	28
B	10	18
C	4	6
D	1	0.4

Table 5.3 Bat Center of Gravity Data

<u>Solution:</u> Since all of the four elements of the bat are symmetrical about the y-axis, the center of gravity of the bat will be on the y-axis. We can therefore define the center of gravity of each element by its position on the y-axis. Table 5.3 gives us the weight and set of coordinates for its center of gravity. Using Eq. 5-4, the center of gravity of the entire bat is calculated as follows:

$$\bar{y} = \frac{\sum_{i=1}^{4} w_i y_i}{\sum_{i=1}^{4} w_i}$$

$$\bar{y} = \frac{(20)(28) + (10)(18) + (4)(6) + (1)(0.4)}{(20) + (10) + (4) + (1)} = 21.8$$

The height of the center of gravity of the bat is 21.8 inches from the end with the cap.

Perhaps no other field gives the center of gravity greater attention than does the field of aircraft design. Airplanes become airborne by generating lift forces on their wings and tail. To keep the plane from rotating in flight, the plane must be designed so that the lift forces are balanced about the center of gravity of the plane. The forces generated by the plane's engines must also be balanced about the center of gravity. The design of the plane's landing gear must consider the position of the center of gravity to maintain stability when the plane is taking off or landing. The cargo of the plane must be positioned in a way which does not drastically change the center of gravity of the plane. If it does change drastically, the pilot will not be able to control the plane. For this same reason, fuel tanks must be positioned so that the plane's center of gravity does not change significantly as the fuel is burned.

The centers of gravity of planes are calculated using the same procedure shown in the bat example. An airplane is symmetrical only about one axis. Aircraft designers usually define the plane's axis of symmetry as being the x-axis. The y-value of the center of gravity will always be 0. To find the location of the center of gravity of a plane, you need to calculate both the x and z values of the center of gravity.

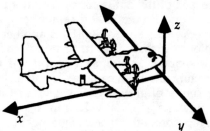

Figure 5.15 Axis Orientaion for Aircraft

Example 5.6: Table 5.4 gives the weight of each major section of a particular airplane. The table also gives the location of the center of gravity of each section of the airplane. Given the data, find the center of gravity of the plane with and without fuel.

Airplane Component	Weight w (pounds)	x Distance (inches)	z Distance (inches)
Wing	235	94.7	71.6
Body/Tail	801	135.8	60.2
Engine/Propeller	545	33.0	44.1
Fuel	390	90.5	71.6
Payload	715	110.2	50.7

Table 5.4 Airplane Center of Gravity

As with the bat problem, it is easiest to organize the calculations with a table. Two columns are added to Table 5.4 to calculate the products of xw and zw. A summation row is added at the bottom. (The easiest way to do this is with an electronic spreadsheet!)

Airplane Component	Weight (pounds)	x Distance (inches)	x•w (inch-pounds)	z Distance (inches)	z•w (inch-pounds)
Wing	235	94.7	22,254.5	71.6	16,826.0
Body/Tail	801	135.8	108,775.8	60.2	48,220.2
Engine/ Propeller	545	33.0	17,985.0	44.1	24,034.5
Fuel	390	90.5	35,295.0	71.6	27,924.0
Payload	715	110.2	78,793.0	50.7	36,250.5
Σ without fuel	2,296		227,808.3		125,331.2
Σ with fuel	2,686		263,103.3		153,255.2

Table 5.5 Center of Gravity Calculations

Using Eq. 5-4, the x and z centers of gravity are calculated as follows:

Without fuel:

$$\bar{x} = \frac{\sum_{i=1}^{4} x_i w_i}{\sum_{i=1}^{4} w_i} = \frac{227,808.3}{2,296} = 99.2" \quad ; \quad \bar{z} = \frac{\sum_{i=1}^{4} z_i w_i}{\sum_{i=1}^{4} w_i} = \frac{125,331.2}{2,296} = 54.6"$$

With fuel:

$$\bar{x} = \frac{\sum_{i=1}^{5} x_i w_i}{\sum_{i=1}^{5} w_i} = \frac{263,103.3}{2,686} = 98.0" \quad ; \quad \bar{z} = \frac{\sum_{i=1}^{5} z_i w_i}{\sum_{i=1}^{5} w_i} = \frac{153,255.2}{2,686} = 57.1"$$

The coordinates of the center of gravity of the plane without fuel are (99.2,0,54.6) with units in inches. The coordinates of the center of gravity with fuel are (98.0,0,57.1). From this, we can see that the center of gravity of the plane will move backward 1.2 in. and down 2.5 in. when the plane goes from full fuel tanks to empty fuel tanks.

Exercise Set 5.4

1. When the mast of a sailboard is
perfectly vertical as shown in Figure
A, the wind-surfer must balance
moments on the sail-board coming
from primarily two forces. The wind
creates a force on the sail which tries
to rotate it in the clockwise direction.
(If the mast arm is vertical, the
weight of the sail has no moment
arm and therefore creates no
moment.) To counter the moment
created by the wind, the wind surfer
must lean backward so that the force

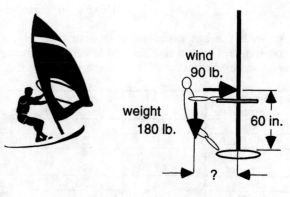

wind
90 lb.

weight
180 lb.

60 in.

?

Figure A

of his weight can create a moment in the counter-clockwise direction. As noted earlier,
the entire weight of an object may be considered to act at its center of gravity. If a
sailboard experiences the forces as shown in Figure A, how far must the windsurfer's
center of gravity be from the mast to counter the forces of the wind?

2. The center of gravity data for each part of a gymnast's body in a half-lever position are
shown in Table A. The distances shown are in inches relative to the bottom of the rings.
The weights are in pounds. The weights of the gymnast's limbs represent only that of one
limb. To determine the location of the center of gravity, the calculations must include the
weights of both limbs.

Figure B

Section	x-Distance	y-Distance	Weight
Head and neck	0	33	11
Trunk	-1	16	77
Upper arm	-1	20	5
Forearm	-1	10	2
Hand	0	1	1
Thigh	6	7	18
Calf	25	6	9
Foot	39	6	2

Table A

a.) Determine the xy coordinates of the gymnast's center of gravity in this position.
b.) As noted earlier, the center of gravity of an object may be considered to be the point at
which the gravitational forces on the object act. Given the gravitational forces that occur
on the gymnast's thighs, calves, and feet, determine the moment created by these parts
on the athlete's waist.

3. An athlete moves across a flat surface by generating lateral forces with his legs. The friction between the athlete's shoes and the ground allows the athlete to move laterally. While the forces move the athlete laterally, they can also cause the athlete's body to rotate about his center of gravity. Figure C shows an infielder moving laterally in two positions, one almost upright and one crouched. In each case, the athlete generates a 20-lb. force to move to his left. The center of

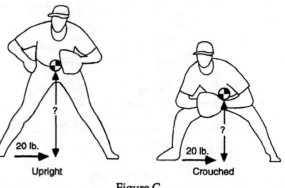

Figure C

gravity data of the athlete's body parts in each position are shown in the Table B. Heights are shown in inches while the weights are shown in pounds. As with the previous problem, the weights shown are for each limb. The center of gravity calculations must include the weights of both limbs. Determine the height of the center of gravity of the athlete and the resulting moment that is created in each position.

Section	Upright	Crouched	Weight
Head and neck	57	46	10.5
Trunk	46	31	72.0
Upper arm	18	32	4.5
Forearm	14	24	2.3
Hand	39	24	0.8
Thigh	29	24	16.5
Calf	12	14	8.3
Foot	2	2	1.5

Table B

4. The part in Figure D is found on a machine. The entire part is made of the same material and has a uniform density. It is made primarily of pieces of 1-cm thick plates. A 2-cm tall cylinder with a cross-sectional area of 3 cm^2 extends 2 cm from each end in the corner. The rest of the dimensions of the part are shown in the figure, also in cm. Determine the coordinates of the center of gravity of the part relative to the origin.

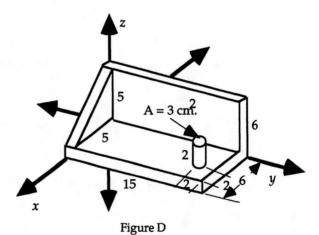

Figure D

5. An airtanker is a plane that is used to drop water on forest fires. The center of gravity data of an airtanker and its water payload are shown in Table C. Determine the difference in the position of the center of gravity with and without the water.

Airplane Component	Weight w (pounds)	x Distance (feet)	z Distance (feet)
Wing	9,315	34.5	12.0
Body/Tail	17,808	35.6	9.3
Engine/ Propeller	4,216	28.0	10.5
Fuel	20,183	34.5	12.0
Payload (water)	26,828	34.0	7.0

Table C

REFERENCES:

E. **Gagnier**, *Inside Gymnastics*, Chicago, Illinois, Henry Regnery Company, 1974. Figure B in Exercise Set 5.4 was adapted from this text.

Chapter 6

SPEED

This chapter discusses the basic relationships between speed, distance, and time. Understanding speed and its relationships with time and distance is necessary to understand the travel of trains, airplanes, missiles, satellites, and all objects that move. As in other chapters, we need to simplify some things to explain the math of travel. This chapter assumes that travel occurs at a constant speed. As you know, travel in sports doesn't usually occur at constant speeds. Sprinters, swimmers, and cyclists need a few seconds to reach their top speed. As the race progresses, athletes tire and their speeds decrease. Near the finish line, the competitors usually try to give their last burst of energy and accelerate to get their best time. Although this chapter studies travel under only constant speed, changing speed, or *acceleration*, will be studied later in the book.

6.1 THE SPEED EQUATION

The fundamental relationship for speed, time, and distance is usually described by the following formula:

$$d = rt \qquad \text{where} \quad d = \text{distance} \tag{6-1}$$
$$r = \text{the } \textit{rate} \text{ of travel, or speed}$$
$$t = \text{time}$$

To help remember the equation, it is often referred to as the "dirt" equation from the pronunciation of the variables.

Example 6.1: If a jogger maintains an average speed of 10 mi./hr., how far can she run in a half hour?

Solution: For $d = rt$ where d = distance, in miles
r = the speed = 10 miles/hour
t = time = 1/2 = 0.5 hr.

Then $d = rt$ = (10 mi./hr.)(0.5 hr.)

= 5 miles

The runner will be able to travel 5 miles in a half hour.

Example 6.2: A cyclist is in a 75 mile race. The record time for the race is 2 hours 30 minutes (2.5 hours). How fast will his average speed need to be to beat this record?

Solution: In this problem, we first need to solve the equation for r.

Again, for $d = rt$ where d = distance = 75 mi.
r = speed, in mi./hr.
t = time = 2.5 hr.

$$\left(\frac{1}{t}\right)d = rt\left(\frac{1}{t}\right)\qquad t = 2.5 \neq 0$$

$$\frac{d}{t} = r$$

$$r = \frac{75 \text{ mi.}}{2.5 \text{ hr.}} = 30 \text{ mi./hr.}$$

The cyclist will need to travel at an average speed faster than 30 mi./hr. to beat the record.

6.1.1 Unit Conversions

In order to do complex travel problems, we often need to be able to convert between different units of distance, speed, and time. One of the most common mistakes in these types of problems is to misuse units. The $d = rt$ formula works only if the units of d, r, and t match. For example, a problem may ask how many *feet* an object will travel if it's traveling at x *miles* per hour. To convert between different units such as feet and miles, we need to use *unit equivalencies*. Some common unit equivalencies are:

Distance	Time
12 inches = 1 foot	60 seconds = 1 minute
3 feet = 1 yard = 36 inches	1 hour = 60 minutes = 3,600 seconds
5,280 feet = 1 mile = 1,760 yards	24 hours = 1 day

3.28 feet = 1 meter
1 mile = 0.54 kilometer

Table 6.1 Unit Equivalencies

The fundamental principle behind unit conversions is the Multiplicative Identity Property. The property says that any real number multiplied by 1 remains the same. The basic idea is that, as long as we multiply the number by these equivalencies of one, the value of the distance, time, or whatever, will remain the same. For example:

12 inches = 1 foot

From the Multiplication Property of Equality, we know that:

$$12 \text{ in.} \left(\frac{1}{1 \text{ ft.}}\right) = 1 \text{ ft.} \left(\frac{1}{1 \text{ ft.}}\right)$$

Therefore $\dfrac{12 \text{ in.}}{1 \text{ ft.}} = 1$

Example 6.3: Ron's fastball is clocked at 90 miles per hour. How fast is this in feet per second?

Solution: In this problem we need to convert both miles to feet and hours to seconds. Using some of the unit equivalencies in Table 6.1, we multiply the 90 miles per hour by equivalencies of one.

$$90 \text{ miles per hour} = \frac{90 \text{ mi.}}{1 \text{ hr.}}$$

$$\frac{5,280 \text{ ft.}}{1 \text{ mi.}} = 1 \qquad\qquad \frac{1 \text{ hr.}}{3,600 \text{ sec.}} = 1$$

$$\frac{90 \text{ mi.}}{1 \text{ hr.}} \left(\frac{5,280 \text{ ft.}}{1 \text{ mi.}}\right)\left(\frac{1 \text{ hr.}}{3,600 \text{ sec.}}\right) = \frac{90(5,280)}{3,600} \text{ ft.}\big/\text{sec.}$$

$$= 132 \text{ ft.}/\text{sec.}$$

Ron's fastball travels at 132 feet per second.

Checking to see that the units cancel out is an extremely important step in any problem like this. Another way to view this is that the units in the numerator and the denominator can be canceled just like numbers.

Exercise Set 6.1

1. The anchor on a 4 x 100m relay team receives the baton and runs at 9 m/s. How far will she travel in 4 seconds?

2. Leena swims a 50m lap in 35 seconds. What was her average speed in meters per second?

3. Gene hits a line drive down the third base line at 162 ft./sec. How long will it take to pass over third base (i.e., travel 90 ft.)?

4. A batter hits a grounder to the short stop. The ball bounces 91 ft. to the infielder at a rate of 100 ft./sec. The batter runs the 90 feet to first base at an average speed of 25 ft./sec. The short stop fields the ball and, a second after catching it, throws it 110 ft. to first base at a speed of 75 ft./sec. Is the batter out?

5. A referee is announcing a penalty to the stands. His announcement travels to a loudspeaker 800 ft. away. The announcement is transmitted to the loudspeaker by radio waves which travel

at nearly the speed of light, 9.84×10^8 ft./sec. The speed of sound is 1,116 ft./sec. How many seconds transpire between the time the ref says something and then hears it over the loudspeaker?

6. A throw is made from a location in left field with the coordinates of (50,190) to second base (90,90). The throw takes 2 seconds to get there. How fast was it thrown in ft./sec.? In mi./hr.?

7. A base runner tries to steal second base. By the time the catcher has the ball and is releasing his throw (above home plate), the base runner is 40 ft. from second and traveling at a speed of 30 ft./sec. How fast will the catcher need to throw the ball to throw out the base runner? (90 ft. between bases)

8. A player on the junior varsity baseball team glances at a varsity player in batting practice at the diamond at the opposite end of the practice field. He notices that there is a half second (0.50) delay between the time when he sees the batter hit a ball and when he hears the hit. The speed of sound is 1,116 ft./sec. Neglecting the time it takes the light to reach him, how far away is the batter?

9. A running back gains 6 yards on a running play. How far is this distance in feet?

10. How many miles does one travel when running a 440 yd. race?

11. A conveyor belt in a factory travels at 3 in./sec. How fast is this in feet per minute?

12. A person who celebrates her 30th birthday has lived how many weeks?

13. A bar on a bench press machine experiences a bending moment of 500 ft.·lb. How many in.·lb. is this?

14. Much of the coast of southern California is traveling northward along the San Andreas Fault. The coast is moving northward at a rate of 2 in./year. For San Francisco and Los Angeles to slide next to each other, the fault will need to move approximately 350 miles. How long will it take before the two cities are next to each other?

15. A city is in the process of setting the timing on one of its traffic signals. It knows that the average walking speed of a pedestrian is 4 ft./sec. With this speed, how long would it take the average pedestrian to cross a street at an intersection that is 60-ft. wide?

6.2 PLOTTING TRAVEL

Plotting travel shows one of the major uses of the Cartesian coordinate system. When travel is plotted, time is usually put on the x-axis and distance is put on the y-axis.

Example 6.4: While running a 10k race, Susan maintains an average speed of 1/3 kilometers per minute (20 km/hr.).

a.) How far will she travel in 0 minutes, 6 minutes, 12 minutes, 21 minutes, and 30 minutes?

b.) Plot her travel on a Cartesian graph.

Solution: a.) Equation 6-1 is used to calculate her distance at each time interval. The values of the distances can be put in a table as follows:

$d = rt = (1/3)t$
$d = (1/3)0 = 0$
$d = (1/3)6 = 2$
$d = (1/3)12 = 4$
etc.

t (min.)	d (km)
0	0
6	2
12	4
21	7
30	10

b.) If we plot these points on a graph with time on the x-axis and distance on the y-axis, we can see that they fall into a straight line.

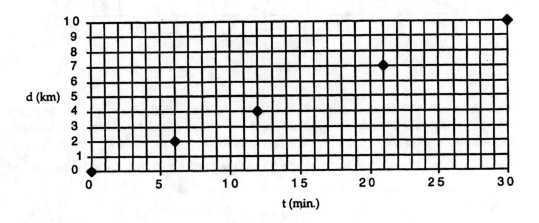

Figure 6.1 Plot of Distances versus Time

If we connect the points with a line, we get a line which represents the formula $d = rt$. Using the xy coordinates with x for time and y for distance, the formula becomes:

$$y = \frac{1}{3}x \qquad \text{where } y = \text{distance, in kilometers}$$

$$x \geq 0 \qquad x = \text{time, in hours}$$

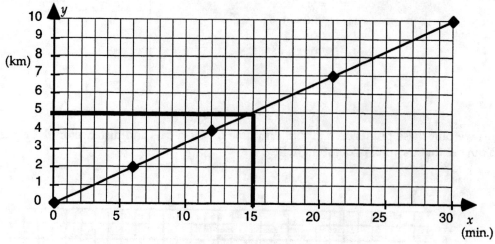

Figure 6.2 Graph of Line $y = \dfrac{1}{3}x$

With this line we can determine the distance she's traveled at any given time. For example, if we want to know her position at 15 minutes, we find 15 minutes on the x-axis and read up to find the point where the line intersects 15 mark. We then look across and read the y value. We can see that it equals 5 km. If we put $x = 15$ into the $y = \dfrac{1}{3}x$ equation, we get the same answer.

$$y = \frac{1}{3}x$$

$$= \frac{1}{3}(15)$$

$$= 5$$

Exercise Set 6.2

Table A gives the travel speeds of some people and vehicles for problems 1-6. The units for each speed are also given. Plot the travel of each person or vehicle as done in Example 6.4. Show the units of each graph on the x- and y-axes. Assume that all travel starts from the point zero (i.e., the lines start at the origin).

Problem	Person/ Vehicle	Speed (slope)	Distance	Time
1	runner	1/4	mi.	min.
2	car	1	mi.	min.
3	cyclist	2/3	mi.	min.
4	hiker	3	mi.	hr.
5	pedestrian	4	ft.	sec.
6	space shuttle	12	mi.	sec.

Table A

Plot the following equations:

7. $y = 2x$ 9. $y = \frac{5}{2}x$ 11. $y = \frac{8}{3}x$ 13. $y = \frac{3}{4}x$ 15. $y = \frac{1}{5}x$

8. $y = \frac{1}{2}x$ 10. $y = \frac{6}{7}x$ 12. $y = \frac{1}{4}x$ 14. $y = 6x$ 16. $y = 0x$

Graph the travel of the people or objects in Table B as done in Exercises 1-6. To make the graphs easier to read, graduate the axes with the increments given in the table having the same length:

Problem	Person/ Object	Speed	Distance	Time	Increment	
					x	y
17.	baseball	90	ft.	sec.	0.1	10
18.	runner	9	yd.	sec.	1	5
19.	hockey puck	150	ft.	sec.	0.1	20

Table B

6.3 SLOPE

A major characteristic of a line is its *slope*. The slope of a line is often described as the "rise over run." In mathematical terms, this is the change in the y-direction divided by the change in the x-direction. The technical definition of the slope, m, of a line which passes through two points (x_1, y_1) and (x_2, y_2) is given by the equation:

$$m = \frac{y_2 - y_1}{x_2 - x_1} = \frac{\Delta y}{\Delta x} \qquad (6\text{-}2)$$

The triangle in the formula, Δ, is delta. Delta is the Greek letter for "d." In math, the Δ represents a *change* in a value. The equation tells us that for any two points on the line between (x_1, y_1) and (x_2, y_2), the ratio of the differences between the x- and y-values remains constant. For Example 6.4, the runner was at 0 at time 0 (0,0) and at 2km at 6 minutes (6,2).

$$\frac{y_2 - y_1}{x_2 - x_1} = \frac{2 - 0}{6 - 0} = \frac{2}{6} = \frac{1}{3}$$

Other points were (21,7) and (30,10)

$$\frac{y_5 - y_4}{x_5 - x_4} = \frac{10 - 7}{30 - 21} = \frac{3}{9} = \frac{1}{3}$$

Note that the slope of the line, $\frac{1}{3}$, equals the runner's speed. Since m is positive on this line, the line is said to have a *positive* slope.

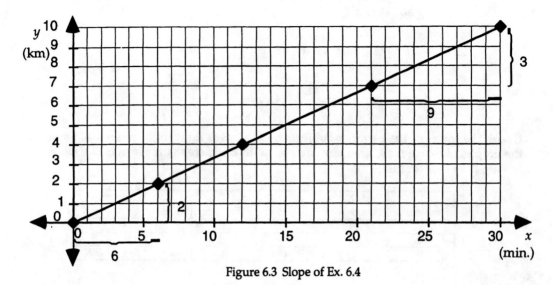

Figure 6.3 Slope of Ex. 6.4

The standard equation for a line passing through the origin is:

$$y = mx. \qquad\qquad (6\text{-}3)$$

The last section showed how to use Eq. 6-2 to determine the slope of a line representing travel at a speed we already knew. A more common use of the equation is to determine the slope from two data points. If the data points represent time and distance, we can use the equation to determine the average speed of the travel.

Example 6.5a: During a 400m race, Andrea maintains a relatively constant speed. At the 100m mark, her time is 10.4 seconds. At the 200m mark, her time is 20.8 seconds. What is her average speed between the 100 and 200m marks?

Solution: The slope can be determined with Eq. 6-2. To use the equation, we need to convert the information into ordered pairs. With time as the x-value and distance as the y-value, the two ordered pairs are (10.4,100) and (20.8,200). Using Eq. 6-2,

$$m = \frac{\Delta y}{\Delta x} = \frac{y_2 - y_1}{x_2 - x_1} = \frac{200 - 100}{20.8 - 10.4} \approx 9.6$$

Her average speed will represent the slope of the line. Her average speed in this interval is therefore 9.6 m/sec.

If we know the slope, we can use Eq. 6-3 to determine her position at any time. Since she maintains a constant speed, the graph of her travel will be a straight line. If we know the slope of the line, we can estimate her locations before, after, or between the two points where she was clocked.

Example 6.5b: With the results of Ex. 6.5a, estimate Andrea's position at 15 and 25 seconds.

Solution: Using Eq. 6-3,

At 15 seconds ($x = 15$) $y = mx = 9.6(15) = 144m$

At 25 seconds ($x = 25$) $y = mx = 9.6(25) = 240m$

At 15 and 25 seconds, her positions are 144m and 240m, respectively.

When you calculate a data point between two known data points (15 is between 10.4 and 20.8), it's called *interpolation*.. When you calculate a point outside of two known points (25 is outside of 10.4 and 20.8) it is called *extrapolation*.

Exercise Set 6.3

Table A shows the times and distances of the travel of some athletes at two different points in a race. They travel at relatively constant speeds. Using Eq. 6-2, determine the speed (slope) of the travel of each person. Given that their travel can be described with the equation $y = mx$, find the missing time or distance at position 3. Show your work.

Problem	Athlete	Units		Position		
		Time (x)	Distance (y)	1	2	3
1	cyclist	hour	miles	(0.5,20)	(2,80)	(3.5,)
2	runner	second	yards	(1.5,12.3)	(4,32.8)	(41,)
3	speed skater	minute	kilometer	(1.8,2)	(3.6,4)	(,3)
4	swimmer	second	meters	(10,15)	(90,135)	(,45)

Table A

Problems 7-11: Determine the slopes of each line in Figure A.

Determine the slopes of the lines that pass through the following points:

10. $P_1(0,2)$; $P_2(-2,-8)$

11. $P_1(8,7)$; $P_2(15,7)$

12. $P_1(4,0)$; $P_2(-4,-7)$

13. $P_1(5,3)$; $P_2(9,-1)$

14. $P_1(-3,-6)$; $P_2(1,-9)$

15. $P_1(11,-20)$; $P_2(51,100)$

16. $P_1(2,-17)$; $P_2(6,-17)$

17. $P_1(-\pi,7)$; $P_2(\sqrt{3},7)$

18. P1(1,0); P2(5,10)

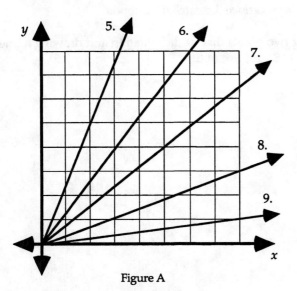

Figure A

6.4 LINES OF THE FORM $y = mx + b$

The equation $y = mx$ is the standard equation of a line passing through the origin. For most travel problems, starting at the origin means starting at time = 0 and the starting position = 0. In sports, competitors often start races at locations other than the zero point. For example in track relays, runners will often receive their batons at different times and at different locations. On the football field, a foot race can start anywhere at any time and travel in any direction.

Example 6.5: Tony catches a pass at his team's 18 yard line and runs down the field at 8 yards per second. Plot his travel. If he's not stopped, how long will it take him to cross the 50 yard line?

Solution: In word problems, it is always helpful to draw a picture to help visualize the problem.

8 yd./sec.

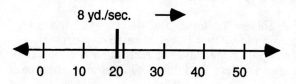

Figure 6.4 Starting Point of Ex. 6.5

As with the previous problems, we'll define time on the x-axis and position on the y-axis. To plot his travel, we need to develop an equation that represents his position as a function of time. Since the problem doesn't give us a starting time, we'll use the time he catches the pass as the starting time ($x = 0$). The distance he'll travel every second after the catch is $8x$ (8 yards/sec. x x seconds). Since he started at the 18 yard line, his position at time x is given by the following equation:

$$y = 8x + 18 \quad \text{where} \quad y = \text{field position, in yards}$$
$$x = \text{time elapsed since the catch, in seconds}$$

Plotting the graph is done by choosing two points that fit the equation and drawing a line through them. We know that he's at the 18 yard line at time = 0 so,

$$y = 18 \quad \text{at} \quad x = 0$$

Therefore, (0, 18) is one point on the line.

At $x = 3$ seconds, $y = 8(3) + 18$

$\qquad\qquad = 24 + 18 = 42$ Therefore the point (3,42) is also on the line.

By plotting the points and connecting them with a line, we get the line which represents the equation.

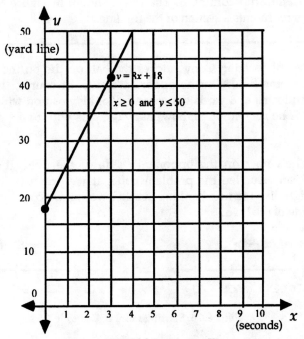

Figure 6.5 Field Position at Time x

To determine when Tony will cross the 50 yard line, we substitute the value of his position at that time, 50, for y in the equation and solve for x.

$$y = 8x + 18$$

$$50 = 8x + 18$$

$$50 - 18 = 8x + 18 - 18$$

$$32 = 8x$$

$$\frac{1}{8}(32) = \frac{1}{8}(8x)$$

$$x = 4$$

If we look at the graph, we can see that Tony crosses the 50 yard line when $x = 4$ sec.

Note that in this example the axes have been graduated differently. The distance of 10 yards in the y-direction is about the same as the distance of 2 seconds in the x-direction. If the axes had the same graduation, the slope of 8 would look much steeper.

It should also be noted that this problem only looks at a particular section of the line $y = 8x + 18$. The graph shows only the values where x is greater than 0 and y is less than 50. The equation, $y = 8x + 18$, also applies to numbers outside of this range. When a line on a graph is not limited by a range, arrows are usually placed on the ends of the lines to show that the relationship continues. The range shown in Figure 6.5 indicates that the line segment only represents a section of the full line.

Example 6.6: Two plays later, Tony catches the ball on the on the 28 yard line on the *opponent's* side of the field. On the opponent's side, of course, the yard markers are decreasing. He still runs at 8 yards per second. Plot his position with zero as the time he catches the ball. If he isn't stopped, how many seconds will it take him to cross the goal line?

Solution: Since he's now on the opponent's side of the field, the yard markers are decreasing. We can visualize the problem better if we orient the yardage markers to increase to the right like the number line. In other words, we could visualize the problem from the other side of the field.

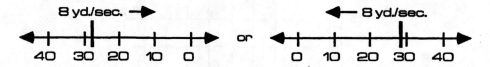

Figure 6.6 Starting Point of Example 6.6

He starts at the 28 yard line. Every second he'll be 8 yards closer to the goal line. His position will therefore decrease by 8 yards per second. His position will be:

$$y = -8x + 28 \quad \text{where} \quad y = \text{field position, in yards}$$
$$x = \text{time elapsed since the catch, in seconds}$$

Again, we can plot a line by finding two points. We know he's at the 28 yard line ($y = 28$) at the start ($x = 0$). So (0,28) is one point.

At $x = 2$ $\qquad y \quad = -8(2) + 28$

$\qquad\qquad\qquad\qquad = -16 + 28 = 12 \qquad$ Plot (2,12)

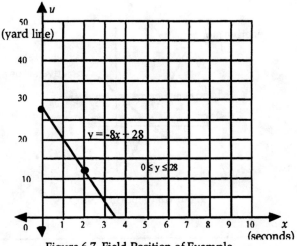

Figure 6.7 Field Position of Example

When Tony crosses the goal line, $y = 0$. So:

$$0 = -8x + 28$$

$$8x + 0 = 8x + (-8x) + 28$$

$$8x = 28$$

$$\left(\frac{1}{8}\right)8x = \left(\frac{1}{8}\right)28$$

$$x = 3.5$$

Tony crosses the goal line 3.5 seconds after catching the ball.

Since m, -8, is negative in this equation, the line $y = -8x + 28$ has a *negative* slope. The examples show that lines which run up to the right and down to the left have a positive slope. Lines which go up to the left and down to the right have a negative slope. For lines which are perfectly horizontal, the y values do not change ($\Delta y = 0$). Therefore:

$$m = \frac{\Delta y}{\Delta x} = \frac{0}{\Delta x} = 0$$

Horizontal lines are said to have a *zero* slope. For vertical lines, the line does not change in the x direction ($\Delta x = 0$). Therefore:

$$m = \frac{\Delta y}{\Delta x} = \frac{\Delta y}{0}$$

The value of m therefore becomes an unreal number. Vertical lines are said to have *no* slope.

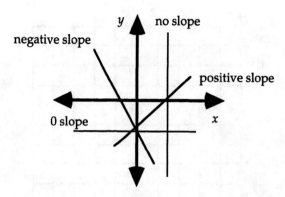

Figure 6.8 Slopes of Lines

Exercise Set 6.4

The table below shows the speeds and starting points of some football players. All travel occurs on the left side of the field. Plot the travel as is shown in Examples 6.5 and 6.6. For each problem, determine when the player crosses the yard line at position A.

Problem	Starting Point yd. marker, b	Speed, m (yd./sec.)	Position A (yd.)
1	12	8	40
2	50	-9	5
3	4	-8	0
4	38	9	11

For the following equations, determine the missing value of x or y.

5. $y = 2x - 5, x = 6$

6. $y = -\dfrac{4}{5}x + 7, x = 9$

7. $y = 0.3x - 4, y = -0.4$

8. $y = -14x + 180, x = 7$

9. $y = 13x + 5, x = 20$

10. $y = 26x + 57, x = 70$

11. $y = -142x - 0.3, x = -7.4$

12. $y = \sqrt{117}x + \pi, y = \pi$

13. $y = \sqrt{5}x - 5, y = 0$

14. $y = mx - b, x = c$

15. $y = mx + b, x = -\dfrac{b}{m}$

16. $y = ax + a, y = 1$

17. $y = -\dfrac{14}{3}x + 2, x = \dfrac{3}{14}$

18. $y = -\sqrt{8}x + 4, y = 0$

19. $y = \sqrt{7}x - 5, x = \sqrt{3}$

20. $y = \sqrt[3]{9}x + 1, y = 4$

Problems 21-25: Determine the slopes of each line in Figure A.

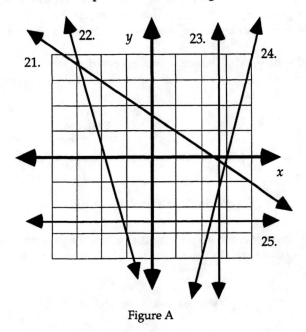

Figure A

6.5 INTERSECTIONS

So far we haven't seen how our receiver gets tackled. On the football field, many players run lines on the field. As any linebacker will tell you, sometimes the lines intersect. Graphs, like football, are much more interesting when the lines intersect.

<u>Example 6.7:</u> Tony catches the ball on the opponent's 46-yard line. He still runs at 8 yards per second. The defensive back covering Tony, Eric, has been "burned" (he let the receiver get past him) and is at the 50 yard line when Tony catches the ball. Eric can run at 9 yards per second. If no other players are involved in the pursuit, will Eric catch up with Tony before the goal line? If so, where does he catch him?

Solution: The first thing we notice from the problem is that the defensive back, Eric, can run faster than the receiver, Tony. From this we can tell that if Eric has enough time, he can catch Tony. We start out by writing the equation for each player's location.

For Tony: $y = -8x + 46$ where y = field position, in yards

                                              x = time elapsed since the catch, in seconds

For Eric: $y = -9x + 50$

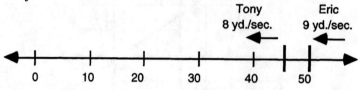

Figure 6.9 Starting Point of Ex. 6.7

The key to solving this problem is knowing that when the players intersect, *they will be at the same place at the same time.* In other words, Eric's location will be the same as Tony's location.

Therefore: $y = -8x + 46 = -9x + 50$

                            $-8x + 46 = -9x + 50$

This is a problem we can solve using our standard algebraic laws.

$$-8x + 46 + 9x = -9x + 50 + 9x$$

$$x + 46 = 50$$

$$x = 4 \text{ seconds}$$

Since we now know when they will intersect, we can determine where they will intersect. We can plug the time value into either Tony's or Eric's equation to determine the position where the intersection occurs.

$$y = -8x + 46$$

$$= -8(4) + 46$$

$$= -32 + 46 = 14 \text{ yards}$$

Double check by substituting 4 into the other equation.

$$y = -9x + 50$$

$$= -9(4) + 50$$

$$= -36 + 50 = 14 \text{ yards}$$

Eric catches up with Tony at the 14 yard line.

If we graph the two lines, we see that the lines intersect at 4 seconds and 14 yards.

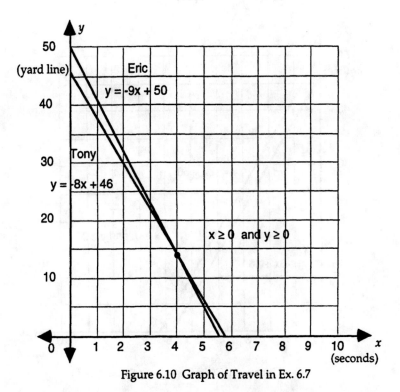

Figure 6.10 Graph of Travel in Ex. 6.7

What would have happened if Eric and Tony ran at the same speed? Eric would have never caught up with Tony. If their speeds were the same, their m values in the equation $y = mx + b$ would be the same. Intuitively we know that if they ran at the same speed, they would stay the same distance apart down the field. If we graph this, we see that the distance between them remains constant as they run downfield. We also notice that the two lines are *parallel*.

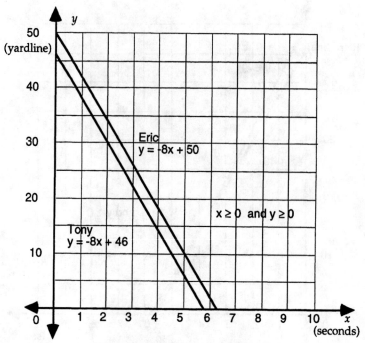

Figure 6.11 Field Positions at Same Speed

Any lines which have the same slope will be parallel.

Example 6.8: Dave is a wide receiver who can run 9 yd./sec and is being covered by Phil who can run 8 yd./sec. A pass play starts at the offense's 6 yard line (yards increasing). Dave catches the pass at the offense's 42 yard line while Phil is at the 39 yard line. Where do the two players intersect?

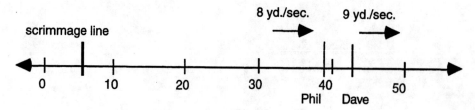

Figure 6.12 Starting Points of Ex. 6.8

<u>Solution:</u> As before, we set up a system of equations:

For Dave $y = 9x + 42$ where y = field position, in yards
                                               x = time elapsed since the catch, in seconds
For Phil $y = 8x + 39$

Again, their positions will be equal when they pass each other. In other words, their y-values will be the same. So,

$$9x + 42 = 8x + 39$$
$$9x - 8x + 42 = 8x - 8x + 39$$
$$x + 42 = 39 - 42$$
$$x = -3$$

The players passed each other at -3 seconds.

What does it mean to say that they passed each other at -3 seconds? The zero point of the time was when Dave caught the ball. If x = -3, they must have passed each other three seconds before Dave caught the ball. We could have suspected this by first looking at the problem. Since Dave was faster and was farther down the field, the only time they could have passed each other was when Dave passed Phil earlier in the play. If we solve one of the equations for y at x = -3, we can see that they did pass each other earlier down the field.

$$y = 9x + 42$$
$$= 9(-3) + 42$$
$$= -27 + 42$$
$$= 15 \text{ yards}$$

The players passed each other at the 15 yard line. The graph of the equations is shown in Figure 6.13.

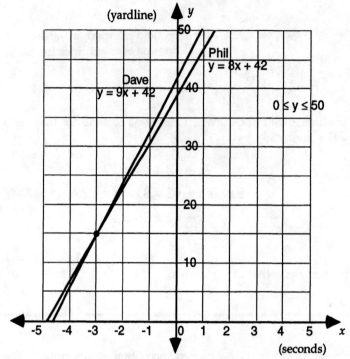

Figure 6.13 Field Positions of Ex. 6.8

In this problem, the intersection of the line occurs in the second quadrant.

Equations which involve the same variables are called *simultaneous equations*. Problems involving simultaneous equations can only be solved if the number of equations equals the number of variables in the problem. In the problems we've done so far, we've had two equations and two variables or *unknowns*.

Even when the number of equations equals the number of variables, a solution may not exist. In the problem where the receiver and the defensive back ran at the same speed, their lines on the graph never intersected. There was no solution. Two simultaneous equations will have no solution if the slopes are the same. Equations with the same slope are said to be *dependent*. When the slopes of two simultaneous equations are different, the equations are said to be *independent*. The solution to two independent equations corresponds to the intersection of their lines on a graph.

Exercise Set 6.5

Table A shows the positions and traveling speeds of some football players in a foot race. As in previous problems, all travel occurs on the left side of the field. All speeds are given in yd./sec. and all positions are given in yards. Determine if the defensive player catches up to the ball carrier and, if so, where.

Problem	Ball Carrier		Defensive Player	
	Speed	Position	Speed	Position
1	--8	39	-9	43
2	-8	46	-8.5	49
3	8	16	8.5	12
4	8.5	24	9	18
5	8.5	36	8	34
6	8	28	8	24

Table A

7. Debbie runs the anchor leg of a 4 x 100m relay. By the time she receives her baton (at 100m from the finish line), her opponent is 3 meters ahead and traveling at 8 m/sec. Debbie runs at 8.5 m/sec. Will she catch her opponent and, if so, where?

Solve the following systems of equations:

8. $\begin{cases} y = 3x - 4 \\ y = 5x + 2 \end{cases}$

9. $\begin{cases} y = 7x + 2 \\ y = 3x - 22 \end{cases}$

10. $\begin{cases} y = -5x + 7 \\ y = 2x + 5 \end{cases}$

11. $\begin{cases} y = 3x + 4 \\ y = 3x - 4 \end{cases}$

12. $\begin{cases} y = 4x + 3 \\ y = 7x - 36 \end{cases}$

13. $\begin{cases} y = \dfrac{1}{3}x - 5 \\ y = \dfrac{2}{3}x + 6 \end{cases}$

14. $\begin{cases} y = 0.3x - 0.1 \\ y = -0.2x \end{cases}$

15. $\begin{cases} y = 4.2x - 6 \\ y = -0.7x + 1 \end{cases}$

16. $\begin{cases} y = \dfrac{1}{\pi}x + 2 \\ y = \dfrac{1}{\pi}x - 3 \end{cases}$

17. $\begin{cases} y = \sqrt{2}x + 4 \\ y = 2\sqrt{2}x + 2 \end{cases}$

18. $\begin{cases} y = -\dfrac{1}{\sqrt{3}}x + 2 \\ y = \dfrac{2}{\sqrt{3}}x + 1 \end{cases}$

19. $\begin{cases} y = \sqrt{5}x + 0.3 \\ y = \sqrt{5}x - 0.1 \end{cases}$

20. $\begin{cases} y = 4x - \sqrt{3} \\ y = 2x + \sqrt{3} \end{cases}$

6.6 THE ADDITION METHOD

The simultaneous problems so far have been solved using a method known as the *substitution method*. With this method, we solved one of the equations for one of the variables and then substituted the value into the other equation. For example in the football problems, we may have solved one of the player's equations for x and then substituted the value of x into the other player's equation.

Along with the substitution method, there are other methods to solve simultaneous equations. The best method to use depends on the type of problem. Another method for solving simultaneous equations is the *Addition Method*. In this method, the equations are lined up, one above the other with the variables and the constant in the same order. One or both of the equations is then multiplied by a non-zero constant and the two equations are then added or subtracted to eliminate one of the variables. The following problem illustrates an example of this method.

Example 6.9: A punt returner catches the ball and starts running 2 yards behind the goal line ($y = $ -2 at time $= 0$). His average speed is 8 yards per second. At the time the punt returner starts running downfield, a player on the punt team passes the 46 yard line at 8 yards per second as shown in Figure 6.14. If the two players maintain these average speeds, where will they meet and at what time? Plot the equations of their travel.

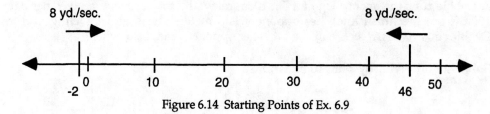

Figure 6.14 Starting Points of Ex. 6.9

Solution: As in the other problems, we'll define the variables as follows:

$x = $ the time elapsed since the punt returner started running, in seconds
$y = $ the field position, in yards

The position of the punt returner can be defined by the equation: $y = 8x - 2$

The position of the punt team player can be defined by the equation: $y = $ -$8x + 46$

Again, the two equations set up a system to solved. When the equations are arranged like this, we notice that if the terms are added vertically, the x-values will disappear.

$$\begin{cases} y = 8x - 2 \\ y = -8x + 46 \end{cases}$$
$$2y = 0 + 44$$

Since we have now eliminated the x, we can solve the resulting equation for y. If both sides of the equation are divided by 2, the result is:

$$y = 22$$

This tells us that the players met at the 22 yard line. To determine the time at which they met, the y value is then substituted back into either of the original equations.

$$y = 8x - 2$$

$$22 = 8x - 2$$

$$24 = 8x$$

$$x = 3$$

This tells us that the players met 3 seconds after the punt returner began running. The two lines may be plotted as shown in Figure 6.15.

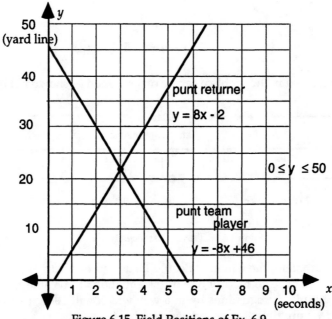

Figure 6.15 Field Positions of Ex. 6.9

The players meet at the 22 yard line, 3 seconds after the punt returner began running.

In most sets of simultaneous equations, it is not so easy to eliminate one of the variables. The coefficients of the variables in the different equations will usually not be the same. When the coefficients are different, we need to multiply one or both of the equations by a non-zero constant that will allow us to eliminate one of the variables.

Example 6.10: Lisa and Sharon are running the two final legs of a 4 x 100 meter relay. Lisa runs the third leg of the race at 8 meters per second and passes the baton to Sharon, the anchor, who runs at 9 meters per second. As usual, track includes 20m take-over zones where the baton is passed. The runners, therefore, may not all carry the baton exactly 100 meters. In this race, the combined distance the two runners travel is exactly 200m. One carries the baton a slightly shorter distance and one a little longer. Their total travel time is 23.5 seconds. Given this information, what is the individual travel time of each runner on her leg of the relay?

Solution: To solve this problem, we'll say

> x = the time Lisa carries the baton, in seconds
> y = the time Sharon carries the baton, in seconds

Since we know their travel speeds, $8x$ = the distance Lisa travels, in meters
                                          $9y$ = the distance Sharon travels, in meters

We also know that the total distance equals 200m and total time equals 23.5 seconds. Our simultaneous equations are therefore:

$$\begin{cases} 8x + 9y = 200 \\ x + y = 23.5 \end{cases}$$

The distances involved in the problem can be diagrammed as shown in Figure 6.16.

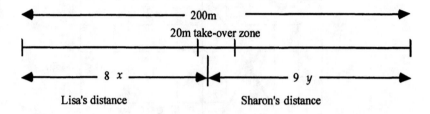

Figure 6.16 Travel Distances in Ex.

In the previous problem, the equations could be added easily to eliminate one variable. In this problem, we need to modify one of the equations to be able to do this. We look for factors to multiply the equations by in a way which will allow us to eliminate one of the variables. If we multiply the second equation by -8,

$$-8 \text{ x} \qquad x + y = 23.5$$

it becomes $\qquad -8x - 8y = -188$

The two equations can now be added vertically:

$$8x + 9y = 200$$

$$\underline{-8x - 8y = -188}$$
$$0x + 1y = 12$$

$$y = 12 \text{ seconds}$$
$$x = 23.5 - 12 = 11.5 \text{ seconds}$$

Lisa carries the baton 12 seconds and Sharon carries it 11.5 seconds.

Exercise Set 6.6

1. With 3 seconds left in a football game, the quarterback throws a pass straight downfield at 20 yd./sec. from the 49 yard line (on the right side of the field). The receiver catches the ball and runs toward the goal line at 9 yd./sec. If the receiver crosses the goal line just as the buzzer goes off, where did he catch the pass?

2. With a base runner on third and no outs, a fly ball is hit to left field. Naturally, the base runner waits until the outfielder catches the ball before running to home. She runs toward home (65 ft.) with an average speed of 20 ft./sec. The outfielder catches the ball at a position 160 ft. from home plate and throws it to her cutoff with a speed of 72 ft./sec. The cutoff, the shortstop, is directly between the outfielder and home. She throws it toward home plate at a speed of 70 ft./sec. It takes each fielder a half second between

catching and throwing the ball. If the ball and base runner arrive at home at the same time, where was the shortstop when she threw the ball?

Solve the following sets of equations using the addition method:

3. $\begin{cases} 2x - 3y = -11 \\ x + 2y = 19 \end{cases}$

6. $\begin{cases} 4x - 3y = 14 \\ 5x - 10y = -20 \end{cases}$

9. $\begin{cases} -x - 3y = -93 \\ 4x - y = 21 \end{cases}$

4. $\begin{cases} x + y = 1 \\ x - y = 3 \end{cases}$

7. $\begin{cases} 7x + 12y = 5 \\ -11x - 24y = -13 \end{cases}$

10. $\begin{cases} 6x - 12y = 24 \\ -3x + 9y = -16 \end{cases}$

5. $\begin{cases} -x + 6y = 21 \\ 7x - 3y = 9 \end{cases}$

8. $\begin{cases} 15x - 4y = 43 \\ x + 2y = 4 \end{cases}$

6.7 OTHER TYPES OF SPEED PROBLEMS

The equations for speed also apply to more practical applications such as the travel of trains, planes, boats, and cars. The following section discusses how the speed equation can be applied to calculate the travel times of airplanes.

 The speed of a plane is divided into different components. The speed which is created by the engines of the plane is known as the *airspeed*. In other words, the airspeed is the speed at which the plane travels through the air. The air, however, may be moving as the plane travels through it; the plane could be traveling through wind. The travel of the plane will therefore be affected by a component known as the *windspeed*. This is the speed of the wind. The airspeed of the plane and the windspeed of the air combine to create the true speed of the plane relative to the ground. The combination of the airspeed and the windspeed is known as the *groundspeed*. In mathematical terms:

$r_g = r_a \pm r_w$ where r_g = speed of plane relative ground, or *groundspeed*

r_a = speed of plane relative to air, or *airspeed*

r_w = speed of the wind, or *windspeed*

It is very important for airlines to know the windspeeds along their routes. The windspeed during a flight will directly affect the fuel consumption of the plane during the flight. Since the fuel represents such a large part of the weight on a flight, it is important not to have too much or too little.

<u>Example 6.11:</u> A plane departs for a city 400 miles away and arrives in an hour and 15 minutes (1.25 hours). During the flight, the plane's instruments indicated that it had an airspeed of 300 miles per hour (mph). How fast and in what direction (headwind or tailwind) was the wind traveling during the flight? (i.e., What was the windspeed during the flight?)

Solution: In this problem, we need to combine the $d = rt$ equation with the groundspeed equation and solve for the windspeed. The speed indicated on the instruments is the airspeed. The speed determined from the distance and travel time is the groundspeed. By knowing these two components, we can determine the windspeed.

$$d = r_g t \qquad\qquad r_g = r_a + r_w$$

$$r_g = \frac{d}{t} = r_a + r_w$$

$$r_w = \frac{d}{t} - r_a$$

$$= \frac{400 \text{ mi.}}{1.25 \text{ hr.}} - 300 \text{ mi./hr.}$$

$$= 20 \text{ mi./hr.}$$

The speed of the wind during the flight was 20 mi./hr. in the direction of the travel of the plane.

The concept of speed can also be used in other applications. For example, rather than calculating how far an object can travel in a given unit of time, a similar equation can be developed to determine how much work a person or machine can do in a given period of time. The following example shows how to the equation can be modified to calculate completed work.

Example 6.12: Al is a mechanic for a small bus agency. His job is to provide routine maintenance for the agency's fleet of 48 buses. If he can service 4 buses a day, how many days will it take to service the entire fleet?

Solution: A formula equivalent to the $d = rt$ equation can be made. We can say that $n = rt$ where

   n = the total number of buses to be serviced
   r = the number of buses serviced per day
   t = the number of days

In this problem, we must solve the equation for t.

$$n = rt$$

$$t = \frac{n}{r} = \frac{48 \text{ buses}}{4 \text{ buses}/\text{day}} = 12 \text{ days}$$

It takes Al 12 working days to service the entire fleet of buses.

The speed problems so far defined the travel rate in terms of distance covered per unit time, $r = d/t$. Similarly, the work problem defined the work rate in terms of units of work per unit of time. Sometimes the rate of work may be defined in terms of units of time required to complete a particular task. For example, the previous problem could have defined the working speed of the mechanic in terms of the amount of time required to service a bus rather than the number of buses serviced per day. The following problem illustrates an example of a work problem with the rate defined in terms of time period per job.

Example 6.13: The janitor of a school can paint the library in 3 days. If the janitor is not available, the library is painted by his assistant. The assistant can paint the library in 5 days. How many days would it take them to paint the library together?

Solution: The problem defines the painting rate of the janitor as 1 job/3 days. The painting rate of the assistant is 1 job/5 days. It is possible to convert work rates of this form to the standard rate of units time per units work. To do this, we simply take the reciprocal of the rate as it is originally given in the project. If the janitor completes the paint job in 3 days, he completes one third of the job per day.

$$r_j = \frac{1}{3} \text{ job} \Big/ \text{day}$$

The amount of work completed by the janitor could be given by the equation:

$$w_j = \frac{1}{3}t \qquad \text{where} \quad w = \text{the amount of work completed,}$$

in jobs
$t = $ the amount of time, in days

Similarly, if the assistant completes the job in 5 days, he completes 1/5 of the job per day.

$$r_a = \frac{1}{5} \text{ job} \Big/ \text{day} \qquad \text{and} \qquad w_a = \frac{1}{5}t$$

If they work together, the amount of work, w_t, completed would be given by the equation:

$$w_t = \frac{1}{3}t + \frac{1}{5}t$$

The objective of the problem is determine the time to complete 1 job; w_t therefore equals 1. Since they would work the same number or hours through completion, the equation can be written as:

$$1 = \left(\frac{1}{3} + \frac{1}{5}\right)t$$

Solving for t:

$$t = \cfrac{1}{\left(\cfrac{1}{3}+\cfrac{1}{5}\right)} = \cfrac{1}{\left(\cfrac{1}{3}\left(\cfrac{5}{5}\right)+\cfrac{1}{5}\left(\cfrac{3}{3}\right)\right)}$$

$$= \cfrac{1}{\left(\cfrac{5}{15}+\cfrac{3}{15}\right)} = \cfrac{1}{\cfrac{8}{15}}$$

$$= \cfrac{15}{8}$$

Working together, it takes the janitor and his assistant 15/8 hours to paint the library.

Exercise Set 6.7

1. Because of a delay, a train is running behind schedule. It has 100 more miles to reach its destination. It is scheduled to arrive in an hour and a 15 minutes (1.25 hours). What will the average speed of the train need to be to reach its destination on time?

2. A soft drink company has a bottling line that can produce 100 bottles of soda per minute.
a.) If run continuously, how many bottles can the line produce in 8 hours?
b.) At this rate, how long would it take to produce 51,000 bottles?
c.) In order to produce 57,600 bottles in the same 8-hour shift, how fast would the line need to run (in bottles/min.)?

3. Newspaper publishers usually prefer to wait as long as possible before printing to make sure that the paper includes the most current information. One newspaper with a circulation of 300,000 has machines which can print its papers in 4 hours. If the newspaper would like to cut its printing time by an hour, what printing speed will its machines need in papers per hour?

4. A pump is used to fill a large cylindrical tank. The tank is 15 feet tall and has a 10-ft. diameter. The pump can fill the tank in 28 minutes. What is the flow rate of the pump in cubic feet per minute?

5. A plane's instruments indicate that it is traveling at 370 miles per hour due east (airspeed). At the time, the plane is experiencing a 40-mile per hour headwind (due west). What is the plane's groundspeed?

6. The radar at an air traffic control center detects that a plane is traveling at 320 mph south (its groundspeed). The air traffic control center also knows that the wind in the area of the plane is 35 mph southward. What airspeed should the plane's instruments indicate?

7. During one hour of a flight, the plane's instruments indicated that it was traveling at 270 mph. Over the hour the plane covered 300 miles.
a.) How fast was the windspeed?
b.) Was it a headwind or a tailwind?

8. To maintain the schedule, a plane must fly 680 miles in 2 hours. The route is experiencing a 35 mph headwind. At what airspeed should the pilot keep the plane to maintain the schedule?

9. To deliver its cargo on schedule, a ship will need to travel 120 miles up a river in 4 hours. The river is flowing at a rate of 3 mph. What will the speed indicators on the ship need to read for the ship to deliver its cargo on time?

10. A church owns two lawn mowers which are used to cut the lawns on its

property. One is a rider and one is a push mower. With the rider, a person can mow the lawns in 1.5 hours. It takes a person 2.5 hours to cut the lawns with the push mower. How many hours would it take two people to cut the lawns using both mowers?

11. The management of a large office building employs two people who wash the building's windows. One of the employees can wash the entire building in 4 days. The other employee requires 6 days to wash the windows. How many days would it take the two to wash the windows together?

12. A dairy owns a machine that can homogenize its regular daily production of milk in five hours. Because of delivery demands, the dairy wants to add another machine to process the milk in three hours. How fast should the new machine be able to process the daily milk production alone?

6.8 BOUNCING

The previous examples showed how travel in one dimension can be monitored relative to time. Graphs may also be used to illustrate two dimensional travel. In the one dimensional travel problems, the slope represented the speed of the travel. In two dimensional travel problems, the concept of slope can be used to indicate the *direction* of the motion. The following section discusses how slopes may be used to indicate direction. To help explain some other important properties of linear equations, this section also introduces some of the science behind bouncing.

Bouncing occurs in several sports. Shooting backboard shots in basketball, shooting pool, and playing racquetball all require an understanding of bouncing. The angle at which an object approaches a surface is known as the *angle of incidence*. The angle at which it bounces is known as the *angle of rebound*. The fundamental principal for bouncing, as shown in Figure 6.17, is that the angle of incidence of the object equals the angle of rebound. The principal also holds true for light reflecting off a mirror. For light, the angle of rebound is known as the angle of *refraction*.

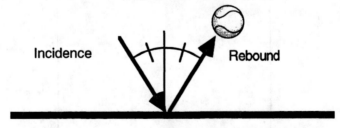

Figure 6.17 Angles of Incidence and Rebound

Other factors can cause the angle of rebound of a bouncing object to be different from the angle of incidence. If an athlete makes the ball spin (or gives it "english"), the angle of incidence will not equal the angle of rebound. If the ball has a top spin, the angle of rebound will be a bit greater than the angle of incidence. If the ball has a back spin, the angle of rebound will be a little less. Friction between the ball and the surface causes the angle of rebound to be a little greater. For this discussion we'll ignore friction.

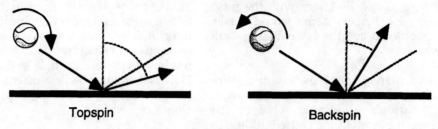

Figure 6.18 Factors Affecting Angle of Rebound

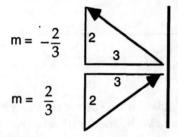

$$m = -\frac{2}{3}$$

$$m = \frac{2}{3}$$

The angle of incidence of a ball could be defined in terms of a slope. In Figure 6.19, the angle of incidence has a slope of $\frac{2}{3}$. If the bouncing occurs in such a way that the angle of incidence equals the angle of rebound, the rebound will have a slope of $-\frac{2}{3}$.

Figure 6.19 Angles by Slope

<u>Example 6.14</u>: Figure 6.20 shows the top view of a pool table with an xy plane superimposed on it. A ball is hit with a path having the equation $y = \frac{4}{5}x - 6$, with units in feet, as shown in the figure. The ball bounces at the end of the table which is located on the line $x = 10$. What is the equation of the path of the ball after it bounces?

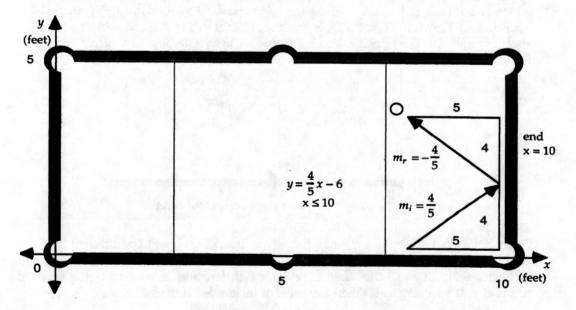

Figure 6.20 Pool Table of Ex. 6.14

Solution: The objective of the problem is to determine the equation of the path of the ball after it hits the end of the table. The line will have the form $y = mx + b$. We therefore need to find the values of m and b. The slope of the line before it hits the end defines the angle of incidence. Since the line has a slope of $m_i = \frac{4}{5}$, the line of the rebound will have a slope of $m_r = -\frac{4}{5}$. The equation will therefore have the form:

$$y = -\frac{4}{5}x + b$$

To complete the equation, we need to determine the value of b. To find b, we need to find one point on the line. After we've found a point on the line, we can substitute the values of the coordinates into the equation and solve the b. We know that one point on the line will be the point at which the ball bounces. To determine the coordinates of that point, we calculate the intersection of the first equation for the path of the ball and the equation of the wall, $x = 10$.

$$\begin{cases} y = \frac{4}{5}x - 6 \\ \quad x = 10 \end{cases}$$

$$y = \frac{4}{5}(10) - 6$$

$$= 4(2) - 6$$

$$= 2$$

The coordinates of the point at which it bounces are therefore (10,2). We can then substitute these coordinates back into the second equation and solve for b.

$$y = -\frac{4}{5}x + b$$

$$2 = -\frac{4}{5}(10) + b$$

$$2 = -4(2) + b$$

$$b = 10$$

The equation of the ball after it bounces is therefore $y = -\frac{4}{5}x + 10$.

When a billiard ball hits a stationary ball, a different type of reaction will occur. If the first ball has no side spin or "english", the two balls will depart from each other at approximately a right angle. The lines which represent their travel will therefore be perpendicular to each other. When two lines are perpendicular, their slopes are *negative reciprocals* of each other.

<u>Example 6.15:</u> A cue ball hits the solid ball and knocks it into the corner pocket as shown in Figure 6.21. The cue ball departs along a path of $y = -\frac{5}{3}x + 17$, with units in feet. The solid ball departs at a right angle with the path of the cue ball. Given that the coordinates of the upper right corner pocket are (10,5),

      a.) What is the equation of the path of the solid ball?
      b.) Does the cue ball scratch?
      c.)

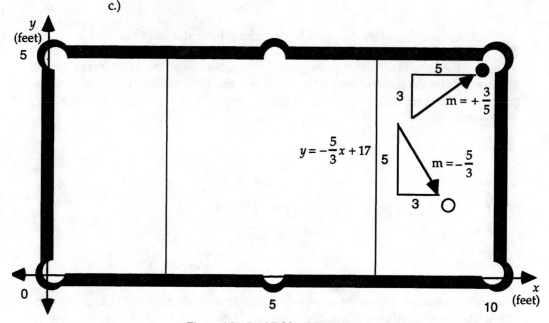

Figure 6.21 Pool Table of Ex. 6.15

<u>Solution:</u> a.) The problem is solved by finding the equation of a line which passes through the point (10,5) and has a slope perpendicular to the path of the cue ball. If the two lines are perpendicular, their slopes are negative reciprocals of each other. If the travel of the solid ball is perpendicular to the line $y = -\frac{5}{3}x + 17$, the slope of the line of its path must be $+\frac{3}{5}$. As in Example 6.14, we find the value of b to fit the equation. We know that the line passes through the point (10,5). To find the value of b, we put the xy values into the equation and solve for b.

$$y = \frac{3}{5}x + b$$

$$5 = \frac{3}{5}(10) + b$$

$$5 = 3(2) + b$$

$$b = 5 - 6 = -1$$

The path of the cue ball is therefore given by the equation with the line:

$$y = \frac{3}{5}x - 1$$

b.) To determine if the cue ball scratches, we see if the path of the ball comes close to the coordinates of the pocket in the direction of the path of the line. In this case, the most likely pocket is the bottom right corner pocket. The coordinates of this pocket are (10,0). One way to solve the problem is to substitute the x-value of the coordinates into the equation and see if the answer equals the y-value of the coordinates. In this case the value of y would have to be very close to 0 for the cue ball to scratch. The equation of the path of the cue ball is:

$$y = -\frac{5}{3}x + 17$$

$$= -\frac{5}{3}(10) + 17$$

$$= -\frac{50}{3} + \frac{51}{3} = \frac{1}{3} \neq 0$$

A third of a foot is 4 inches which is not close enough to scratch.

Exercise Set 6.8

1. On a putting green, the ball is hit in a level area toward the hole on a path with the equation $y = -1.5x + 7$. The coordinates of the hole are (8,-5). Is the put good?

2. A hockey player makes a shot sending the puck on a path described by the line $y = -5x + 285$. The posts of the goal are at (47,10) and (53,10). If it's not blocked, will the shot be good?

Problems 3 through 6 involve some examples of objects bouncing off surfaces. Assume that the angle of incidence equals the angle of rebound in all of the problems. Where two spheres collide, assume that they depart at right angles.

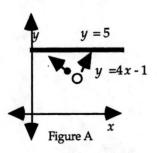

Figure A

3. A cue ball hits the 3 ball and then departs on a path with the equation $y = 4x - 1$ as shown in Figure A. The second ball then hits a side positioned on the line $y = 5$. It then hits another side whose position is $x = 0$. What slope will the path of the ball have after it hits the second wall?

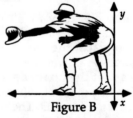

Figure B

4. After fielding a grounder, a shortstop throws the ball to first. The ball, however, is thrown low requiring the firstbaseman to dig it out. Just before the ball hits the ground, it is traveling on a path given by the equation:

$$y = -\frac{1}{5}x - 24$$

where y is the height of the ball and x is the horizontal distance from first base, both with units in inches. The first-baseman can extend his arm and glove a length of 80 inches ($x = -80$) as shown in Figure B. Assuming that the travel of the ball after it bounces remains linear, how high above the ground (y = ?) should he place his glove to trap the ball?

5. A cue ball hits the 5-ball located at the coordinates (2,2) and then departs along a path of:

$$y = \frac{2}{3}x + 1$$

a.) What is the equation of the resulting path of the 5-ball?
b.) Does the 5-ball make it into the pocket with the coordinates (0,5)?

6.

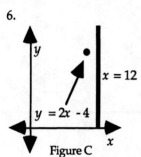

Figure C

6. A racquetball player hits the ball toward the wall on a path with the equation $y = 2x - 4$ as viewed from above. The position of the wall is given by the equation $x = 12$ as shown in Figure C. What equation describes the path of the ball after it hits the wall?

7. The pump on a city's well used to be run with a diesel engine. The city now wants to replace the engine with an electric motor. A power line adjacent to a road passes near the pump on a path with the equation of $y = 3x - 1,460$ with units in feet as shown in Figure D. The coordinates of the pump are (60,200) with units in feet. What is the least amount of wiring necessary to provide electrical service from the power line to the pump?

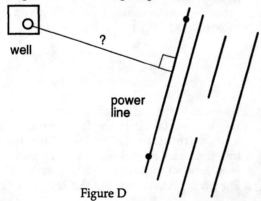

Figure D

8. A family would like to add another bedroom to their home as shown in Figure E. Currently, the back wall of the home is located on a line given by the equation $y = 2x - 8$. Because of city ordinances, they cannot extend their house any closer than 4 feet from the fence (at $y = 0$). The new corner of the house extends as far as possible ($y = 4$) and remains at a right angle. What equation will describe the line on which the new side wall is located ?

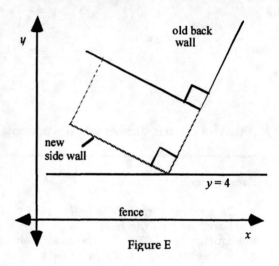

Figure E

Chapter 7

SYSTEMS OF EQUATIONS

In Chapters 2 and 3, we learned some of the rules and procedures to solve equations with one variable. Chapter 6 introduced some methods of solving systems with 2 equations and 2 unknowns. Problems can also occur which involve three or more equations and unknowns. This chapter discusses two methods for solving problems with three or more equations and unknowns.

Problems which involve large systems of equations require some rather complicated procedures to solve. To make the large systems easier to work with, the variables are removed from the coefficients. The processes involve arranging the coefficients and the constants into arrays of numbers called *matrices*. From the system,

$$\begin{cases} 2x + 3y = -5 \\ 7x - 4y = 1 \end{cases}$$

the following matrix would be formed:

$$\begin{vmatrix} 2 & 3 & -5 \\ 7 & -4 & 1 \end{vmatrix}$$

By arranging the numbers into orderly rows and columns it is easier to apply the rather sophisticated procedures which are used to solve the systems.

7.1 GAUSSIAN ELIMINATION

A German mathematician by the name of Karl Friedrich Gauss expanded the Addition Method (learned in the last chapter) into a sophisticated mathematical process. The goal of Gaussian elimination is to convert the matrix to the form:

$$\begin{bmatrix} 1 & - & - & - \\ 0 & 1 & - & - \\ 0 & 0 & 1 & - \end{bmatrix}$$

Recall that when we used the Addition Method, we tried to get the equation down to the form $y = 3$, for example. Another way to describe this equation is $1y = 3$. Still another way to write it is $0x + 1y = 3$. This is why each of the lower rows in the matrix above has a 1 preceded by zeroes.

In order to make the conversion, we use the following rules:

- Rows can be exchanged

- The columns of coefficients (not the column of constants) can be exchanged
- Rows can be multiplied or divided by a non-zero constant
- Rows can be added and subtracted to/from each other

Although the rows can be changed, you need to keep track of which column represents which variable.

The following example shows how a matrix of numbers is developed and processed.

<u>Example 7.1:</u> In a basketball game, Dave scores a total of 29 points. He scores the 29 points with a combination of 2-point and 3-point baskets and free throws. In the game, he makes a total of 15 *baskets* (including all shots). He makes twice as many 2-point shots as he does 3-point shots and free throws combined. How many baskets does he make of each type?

<u>Solution:</u> As always, the first step is to define the variables.

a = the number of 1-point baskets (free throws)
b = the number of 2-point baskets
c = the number of 3-point baskets

The next step is to set up the system of equations. Since there are 3 variables in the problem and the problem gives us 3 independent relationships between the variables, we have enough information to solve the problem.

Since Dave scores a total of 29 points:

$$1a + 2b + 3c = 29$$

Since he makes a total of 15 baskets:

$$a + b + c = 15$$

If he scores twice as many 2-point shots as he does 3-point shots and free throws combined:

$$b = 2(a + c)$$

Gaussian elimination requires setting up the equations with the variables and the constant in the same order. The first two equations have the variables and the constant in the same order. The variables in the third equation are in a different order. We need to rearrange the variables and the constant in the third equation to get them into the same order as the first two equations.

$$b = 2(a + c)$$

$$b = 2a + 2c$$

$$b - b = 2a - b + 2c$$

$$2a - b + 2c = 0$$

Therefore, our system of equations is:

$$\begin{cases} a + 2b + 3c = 29 \\ a + b + c = 15 \\ 2a - b + 2c = 0 \end{cases}$$

We begin the elimination by setting up a matrix of only the coefficients of the variables and the constants. For this problem, the matrix is:

$$\begin{bmatrix} 1 & 2 & 3 & 29 \\ 1 & 1 & 1 & 15 \\ 2 & -1 & 2 & 0 \end{bmatrix}$$

Again, the goal of Gaussian elimination is to convert the matrix to the form:

$$\begin{bmatrix} 1 & - & - & - \\ 0 & 1 & - & - \\ 0 & 0 & 1 & - \end{bmatrix}$$

Since a "1" already exists in the top left corner of our matrix, there's no need to exchange rows or columns to put one there. The first row can remain as it is. The next step is to get the number in the a column in the second row to be a 0 and the number in the b column to be a 1. We notice that this can be achieved by subtracting the second row from the first and replacing the second row with this new result. The resulting matrix is:

$$\begin{bmatrix} 1 & 2 & 3 & 29 \\ 0 & 1 & 2 & 14 \\ 2 & -1 & 2 & 0 \end{bmatrix}$$

We now need to get the a and b columns of the third row to be 0 and the c column of the third row to be 1. First we get rid of the 2 in the first row. We can do this if we multiply the first row by 2 and subtract the third row from it.

Row 1 x 2:
$$\begin{bmatrix} 2 & 4 & 6 & 58 \\ 0 & 1 & 2 & 14 \\ 2 & -1 & 2 & 0 \end{bmatrix}$$
Row 1 - Row 3:
$$\begin{bmatrix} 2 & 4 & 6 & 58 \\ 0 & 1 & 2 & 14 \\ 0 & 5 & 4 & 58 \end{bmatrix}$$

We can change the first row back to its original form.

$$\begin{bmatrix} 1 & 2 & 3 & 29 \\ 0 & 1 & 2 & 14 \\ 0 & 5 & 4 & 58 \end{bmatrix}$$

We now need to get the 5 to be a 0 and the 4 to be a 1. We multiply the second row by 5 and subtract the third row from it.

$$\text{Row 2 x 5:} \quad \begin{bmatrix} 1 & 2 & 3 & 29 \\ 0 & 5 & 10 & 70 \\ 0 & 5 & 4 & 58 \end{bmatrix} \qquad \text{Row 2 - Row 3:} \quad \begin{bmatrix} 1 & 2 & 3 & 29 \\ 0 & 5 & 10 & 70 \\ 0 & 0 & 6 & 12 \end{bmatrix}$$

When the second row is converted back to its previous form, the matrix is:

$$\begin{bmatrix} 1 & 2 & 3 & 29 \\ 0 & 1 & 2 & 14 \\ 0 & 0 & 6 & 12 \end{bmatrix}$$

To get the 6 in the third row to be a 1, we simply divide the row by 6 (multiply by 1/6). The final matrix is:

$$\begin{bmatrix} 1 & 2 & 3 & 29 \\ 0 & 1 & 2 & 14 \\ 0 & 0 & 1 & 2 \end{bmatrix}$$

which has the form we were looking for.

Once we have the matrix in this form, the equations are very easy to solve. From the third row we see that:

$$0a + 0b + 1c = 2 \qquad c = 2$$

The answer for c is then substituted into the second row.

$$0a + 1b + 2(2) = 14 \qquad b = 14 - 4 = 10$$

Then the values for b and c are substituted into the first row.

$$1a + 2(10) + 3(2) = 29 \qquad a = 29 - 20 - 6 = 3$$

Dave makes 3 free throws, 10 2-point baskets, and 2 3-point baskets.

Exercise Set 7.1

1. During a football game, the Tigers score 10 times for a total of 48 points by touchdowns with PATs (7 points each), field goals (3 points each), and safeties (2 points each). The number of touchdowns scored by the team equals the sum of the number of 3-point field goals and safeties combined. How many of each score did the team make?

Use Gaussian elimination to solve the following systems of equations:

2. $\begin{cases} 3x + 7y - 4z = 17 \\ -x - y + 2z = -5 \\ 9x + 5y - 6z = 29 \end{cases}$

3. $\begin{cases} 2x - y + z = -8 \\ 5x + 2y - z = 13 \\ -7x + 4y + 3z = 11 \end{cases}$

$$4. \begin{cases} -6x + 7y + z = 19 \\ -4x - 2y - 10z = -46 \\ 3x - 3y + 8z = 18 \end{cases}$$

$$6. \begin{cases} 3a + 2b + c = 9 \\ a + 2b + 4c = 13 \\ 4a + 4b + 6c = 24 \end{cases}$$

$$5. \begin{cases} 2x + 6y + 2z = 46 \\ 3x + 2y + 5z = 18 \\ -2x + 1y + 4z = -3 \end{cases}$$

$$7. \begin{cases} 9x - y + 2z = 25 \\ 2x + 2y + 2z = 8 \\ -x + 5y + 3z = 2 \end{cases}$$

7.2 CRAMER'S RULE

Another method of solving systems is to use Cramer's Rule of determinants. Although this method can take a little longer, it is always a straightforward process. It doesn't require the planning used in Gaussian elimination. The Gaussian elimination process can also be more difficult when the coefficients are decimals.

With Cramer's Rule, a matrix is also formed by isolating the coefficients of the variables and the constants. Sets of *determinants* are then formed from the matrix. For the matrix:

$$\begin{bmatrix} a_1x & b_1y & c_1 \\ a_2x & b_2y & c_2 \end{bmatrix}$$

The following determinants are formed:

$$D_x = \begin{vmatrix} c_1 & b_1 \\ c_2 & b_2 \end{vmatrix} \qquad D_y = \begin{vmatrix} a_1 & c_1 \\ a_2 & c_2 \end{vmatrix} \qquad D = \begin{vmatrix} a_1 & b_1 \\ a_2 & b_2 \end{vmatrix}$$

Solving determinants involves a special mathematical process. The value of a 2 by 2 determinant is calculated using the following equation:

$$\begin{vmatrix} a_1 & b_1 \\ a_2 & b_2 \end{vmatrix} = a_1b_2 - a_2b_1 \qquad (7\text{-}1)$$

Cramer's Rule states that for the three determinants:

$$x = \frac{D_x}{D} \qquad y = \frac{D_y}{D} \qquad (7\text{-}2)$$

Since using Cramer's rule requires finding the value of the D determinant, which is in the denominator in the equations above. Cramer's rule is only valid when $D \neq 0$.

7.2.1 Practical Applications

Students often find it helpful to understand how simultaneous equations and matrices occur in the professional world. To illustrate one example, the following problem shows how determinants are used to solve problems that are routinely faced by designers of airplanes. This kind of problem helps explain why determinants are necessary. To better appreciate the problem, some background on aircraft design is provided. While the discussion may be helpful to some students, it is not required to understand Cramer's Rule. It is possible for the reader to skip the discussion and proceed to the next example problem.

Two of the most important characteristics of an airplane are its *lift* and its *drag*. The lift force of a plane represents its ability to lift the plane and any cargo. The more lift the plane can generate, the better. The drag force of an airplane is similar to the wind resistance of a car. The more wind resistance a car has, the worse gas mileage it will get. The same is true for planes. Unfortunately, in order to get airborne, a plane will need at least some wind resistance. Part of the energy lost in wind resistance is necessary to lift the plane. In other words, the lift force of a plane creates some drag. The lift and drag of a plane therefore depend on each other.

The actual lift and drag of a plane depend on the speed and altitude of the plane. A plane resting on the ground has zero lift force but while traveling at several hundred miles per hour it may be able to lift thousands of pounds. At higher altitudes, the air is thinner which makes it harder to lift the plane. Rather than having absolute lift and drag forces, a given plane will have lift and drag *coefficients*. Once these coefficients are known, designers can determine how the aircraft will perform at any speed and at any altitude.

One common way of determining the lift and drag coefficients of an airplane is to run tests of a model of the plane in a wind tunnel. The wind tunnel creates a condition of wind traveling past the plane which simulates conditions of the plane in flight. As the wind tunnel is forcing air by the plane, designers take measurements with instruments which help them determine the lift and drag forces on the plane under different velocities. The plane model can also be positioned at different angles. This allows the engineers to see how the plane will perform at different angles. By determining the lift and drag forces under different conditions, the engineers can develop equations that can be solved to determine the lift and drag coefficients.

In the following problem, the two variables are K_1 and K_2. K_1 is associated with the wind resistance drag ; it is the *parasite* drag coefficient. K_2 is associated with the drag created by lift; it is known as the *induced* drag coefficient.

Example 7.2: An aircraft model is mounted in a wind tunnel for aerodynamic study. The lift and drag forces of the model are measured at two different flight conditions. After the tests are conducted and some calculations are performed on the data, the following system of equations is developed:

$$\begin{cases} K_1 + 0.36K_2 = 0.0296 \\ K_1 + 1.44K_2 = 0.0674 \end{cases}$$

Using determinants, determine the parasite and induced drag coefficients K_1 and K_2.

Solution: The system in this problem represents a 2 x 2 matrix, that is two equations and two unknowns. To solve a 2 x 2 equation using matrices and Cramer's Rule, we set up three determinants from the system.

$$D = \begin{vmatrix} 1 & 0.36 \\ 1 & 1.44 \end{vmatrix} \qquad D_{K_1} = \begin{vmatrix} 0.0296 & 0.36 \\ 0.0674 & 1.44 \end{vmatrix} \qquad D_{K_2} = \begin{vmatrix} 1 & 0.0296 \\ 1 & 0.0674 \end{vmatrix}$$

Using Eq. 7-1, we determine the value of each determinant.

$$D = \begin{vmatrix} 1 & 0.36 \\ 1 & 1.44 \end{vmatrix} = 1(1.44) - 1(0.36) = 1.08$$

$$D_{K_1} = \begin{vmatrix} 0.0296 & 0.36 \\ 0.0674 & 1.44 \end{vmatrix} = 0.0296(1.44) - 0.0674(0.36) = 0.01836$$

$$D_{K_2} = \begin{vmatrix} 1 & 0.0296 \\ 1 & 0.0674 \end{vmatrix} = 1(0.0674) - 1(0.0296) = 0.0378$$

From Equation 7-2: $K_1 = \dfrac{D_{K_1}}{D} = \dfrac{0.01836}{1.08} = 0.017 \quad K_2 = \dfrac{D_{K_2}}{D} = \dfrac{0.0378}{1.08} = 0.035$

From the wind tunnel tests, the values of K_1 and K_2 are determined to be 0.017 and is 0.035 respectively.

7.2.2 Three by Three Matrices

Determinants of virtually any size can be solved using Cramer's rule. The only requirement is that the number of independent equations equals the number of unknowns. For systems with more than two variables, an additional process must be added to that used for 2 x 2 matrices. For the following 3 x 3 matrix:

$$\begin{bmatrix} a_1x & b_1y & c_1z & d_1 \\ a_2x & b_2y & c_2z & d_2 \\ a_3x & b_3y & c_3z & d_3 \end{bmatrix}$$

The following determinants are formed:

$$D_x = \begin{vmatrix} d_1 & b_1 & c_1 \\ d_2 & b_2 & c_2 \\ d_3 & b_3 & c_3 \end{vmatrix} \quad D_y = \begin{vmatrix} a_1 & d_1 & c_1 \\ a_2 & d_2 & c_2 \\ a_3 & d_3 & c_3 \end{vmatrix} \quad D_z = \begin{vmatrix} a_1 & b_1 & d_1 \\ a_2 & b_2 & d_2 \\ a_3 & b_3 & d_3 \end{vmatrix} \quad D = \begin{vmatrix} a_1 & b_1 & c_1 \\ a_2 & b_2 & c_2 \\ a_3 & b_3 & c_3 \end{vmatrix}$$

As with 2 x 2 matrices, Cramer's Rule states that for the determinants above:

$$x = \frac{D_x}{D} \qquad y = \frac{D_y}{D} \qquad z = \frac{D_z}{D}$$

The value of a 3 x 3 matrix is calculated by multiplying the values in the top row by their respective *minors*. The minors in a 3 x 3 matrix are 2 x 2 matrices. The minors of the three terms in the top row are the sets of numbers not crossed out as shown in the matrices below:

$$\begin{vmatrix} a_1 & b_1 & c_1 \\ a_2 & b_2 & c_2 \\ a_3 & b_3 & c_3 \end{vmatrix} \qquad \begin{vmatrix} a_1 & b_1 & c_1 \\ a_2 & b_2 & c_2 \\ a_3 & b_3 & c_3 \end{vmatrix} \qquad \begin{vmatrix} a_1 & b_1 & c_1 \\ a_2 & b_2 & c_2 \\ a_3 & b_3 & c_3 \end{vmatrix}$$

The process works out to the following equation:

$$\begin{vmatrix} a_1 & b_1 & c_1 \\ a_2 & b_2 & c_2 \\ a_3 & b_3 & c_3 \end{vmatrix} = a_1 \begin{vmatrix} b_2 & c_2 \\ b_3 & c_3 \end{vmatrix} - b_1 \begin{vmatrix} a_2 & c_2 \\ a_3 & c_3 \end{vmatrix} + c_1 \begin{vmatrix} a_2 & b_2 \\ a_3 & b_3 \end{vmatrix} \qquad (7\text{-}3)$$

Note that in Equation 7-3, signs of the top row of coefficients alternate between positive and negative a_1 is positive, b_1 is negative, and c_1 is positive. The pattern continues regardless of the size of the matrix. The 2 by 2 determinants are then solved using Equation 7-1.

Example 7.3: The following system of equations arises when designing an electronic circuit. The I values represent the electrical current passing through the components 1, 2, and 3. Electrical current is measured in units called *amperes*. Using determinants, solve the system for the values of the currents, I_1, I_2, and I_3.

$$\begin{cases} 4I_1 - 2I_2 + 3I_3 = 5 \\ -5I_1 + 4I_2 - 4I_3 = 4 \\ -I_1 + 3I_2 - 8I_3 = 9 \end{cases}$$

Solution: The problem is begun by setting up the determinants as follows. The calculation of the determinants here will be progressively simplified to illustrate the steps which can be done in your head.

$$D_{I_1} = \begin{vmatrix} 5 & -2 & 3 \\ 4 & 4 & -4 \\ 9 & 3 & -8 \end{vmatrix} = 5 \begin{vmatrix} 4 & -4 \\ 3 & -8 \end{vmatrix} - (-2) \begin{vmatrix} 4 & -4 \\ 9 & -8 \end{vmatrix} + 3 \begin{vmatrix} 4 & 4 \\ 9 & 3 \end{vmatrix}$$

$$= 5[(4)(-8) - (3)(-4)] + 2[(4)(-8) - (9)(-4)] + 3[(4)(3) - (9)(4)]$$

$$= 5(-32 + 12) + 2(-32 + 36) + 3(12 - 36)$$

$$= 5(-20) + 2(4) + 3(-24)$$

$$= -100 + 8 - 72$$

$$= -164$$

$$D_{I_2} = \begin{vmatrix} 4 & 5 & 3 \\ -5 & 4 & -4 \\ -1 & 9 & -8 \end{vmatrix} = 4 \begin{vmatrix} 4 & -4 \\ 9 & -8 \end{vmatrix} - 5 \begin{vmatrix} -5 & -4 \\ -1 & -8 \end{vmatrix} + 3 \begin{vmatrix} -5 & 4 \\ -1 & 9 \end{vmatrix}$$

$$= 4[(4)(-8) - (9)(-4)] - 5[(-5)(-8) - (-1)(-4)] + 3[(-5)(9) - (-1)(4)]$$

$$= 4(-32 + 36) - 5(40 + 4) + 3(-45 + 4)$$

$$= 16 - 180 - 123$$

$$= -287$$

$$D_{I_3} = \begin{vmatrix} 4 & -2 & 5 \\ -5 & 4 & 4 \\ -1 & 3 & 9 \end{vmatrix} = 4(36 - 12) - (-2)(-45 - (-4)) + 5(-15 - (-4))$$

$$= 4(24) + 2(-41) + 5(-11)$$

$$= 96 - 82 - 55$$

$$= -41$$

$$D = \begin{vmatrix} 4 & -2 & 3 \\ -5 & 4 & -4 \\ -1 & 3 & -8 \end{vmatrix} = 4(-32 - (-12)) - (-2)(40 - 4) + 3(-15 - (-4))$$

$$= 4(-20) + 2(36) + 3(-11)$$

$$= -80 + 72 - 33$$

$$= -41$$

$$I_1 = \frac{D_{I_1}}{D} = \frac{-164}{-41} = 4 \qquad I_2 = \frac{D_{I_2}}{D} = \frac{-287}{-41} = 7 \qquad I_3 = \frac{D_{I_3}}{D} = \frac{-41}{-41} = 1$$

The currents passing through components 1, 2, and 3 are 4, 7, and 1 amperes, respectively.

In some 3 x 3 problems, it may be easier to calculate the determinants of two of the variables, and then substitute the values into one of the original equations to calculate the value of the third variable.

Exercise Set 7.2

1. In a 4 x 100m relay race, the first three runners in the race run at speeds of 9.5, 10.0, and 10.2 m/sec. Each runner carries the baton a different length. The total distance the three runners carry the baton is 306m. The distance the third runner carries the baton is the average distance of the other two runners. The total time the three runners carry the baton is 30.9 seconds. Use Cramer's Rule to determine the amount of time each runner carries the baton.

Use Cramer's rule to solve the following systems of equations:

2. $\begin{cases} 3x + 7y - 4z = 17 \\ -x - y + 2z = -5 \\ 9x + 5y - 6z = 29 \end{cases}$

6. $\begin{cases} 6x + 1y + 5z = 6.3 \\ -1x - 2y + 4z = -1 \\ 2x + 3y + 3z = 3.7 \end{cases}$

3. $\begin{cases} 2x - y + z = -8 \\ 5x + 2y - z = 13 \\ -7x + 4y + 3z = 11 \end{cases}$

7. $\begin{cases} 9x - y + 2z = 25 \\ 2x + 2y + 2z = 8 \\ -x + 5y + 3z = 2 \end{cases}$

4. $\begin{cases} -6x + 7y + z = 19 \\ -4x - 2y - 10z = -46 \\ 3x - 3y + 8z = 18 \end{cases}$

8. $\begin{cases} 5x - 2y - 4z = 0 \\ -3x + 7y + z = 29 \\ 2x - 8y - 9z = -41 \end{cases}$

5. $\begin{cases} -4x + 2y + 6z = 10.8 \\ 10x + 1y + 5z = -9 \\ 8x + 4y - 1z = 2.4 \end{cases}$

Chapter 8

VECTORS

This chapter continues the study of motion. Like most of the other subjects in this book, motion is better understood when it is broken down into components. The motion of any object can be broken down into its *translations* (or *displacements*) and its *rotation*. For example, a pass in football involves a displacement from the passer to the receiver. While the ball is moving from passer to receiver, it is also rotating on an axis. Its displacement and rotation can be defined by the use of *vectors*. Vectors are numerical terms which indicate the magnitude and direction of the motion.

Vectors can be used to describe the motion of a simple object or a complex piece of machinery. For example, the motion of a pitcher in baseball can be described in terms of translation and rotation vectors. Figure 8.1 illustrates the complicated motion of pitching a baseball. Beginning from the ground, the pitcher's back leg pushes against the rubber moving the pitcher's upper body toward the batter. This is a translation. Most of the remainder of the motion involves rotations. The pitcher's upper body (or torso) rotates about the hips. His arm rotates about his shoulder. His forearm rotates about his elbow and his hand rotates about his wrist.

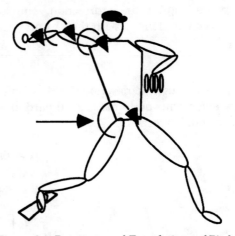

Figure 8.1 Rotations and Translations of Pitching

Vectors are unique because they always define the direction of the motion as well as its magnitude. In Chapter 6, we described speed by saying the receiver ran at 8 yards per second. Speed only describes the *magnitude* of the motion. It is therefore called a *scalar quantity*. To say the receiver traveled at 8 yards per second *straight downfield* also describes the direction of the motion. When direction is included with the speed it is called *velocity*. Velocity is a *vector quantity*.

If a swimmer swims one full lap (down and back) in a 50 meter pool in 1 minute, her average *speed* is:

$$\frac{50 + 50}{1} = \frac{100 \text{ m}}{1 \text{ min.}} = 100 \text{ }^{m}\!/_{min.}$$

When using vector quantities, the direction of her travel is considered. We consider her motion as if it occurred on an xy plane. A 50 m displacement in the positive x-direction is described by the term $50i$. The 50 m displacement on the way back occurs in the opposite direction. It is therefore described as $-50i$. Her average *velocity* is therefore:

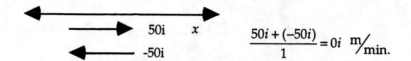

$$50i \quad x \qquad \frac{50i + (-50i)}{1} = 0i \ ^m\!/_{min}.$$

$$-50i$$

8.1 DISPLACEMENT VECTORS

Motion in the xyz coordinate system is defined by *unit vectors*. The unit vectors *ijk* are used to identify motion in the $x, y,$ and z directions respectively. For example, an object that moves 1 unit in the x direction, 2 units in the y direction, and 3 units in the z-direction can be defined with the *ijk* unit vectors as:

$$1i + 2j + 3k$$

Vectors can be used to describe the down-and-out pass pattern discussed in Chapter 5. The receiver who performs the 10-yard down and 5-yard out does a pattern with a displacement vector **A** of:

$$A = 10i + 5j$$

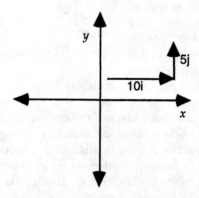

Figure 8.2 Displacement Vectors

Vectors do not need to start or finish at the same point to be equal. The vectors **A** and **B** in Figure 8.3 have the same magnitudes and directions and are therefore equal.

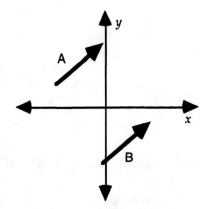

Figure 8.3 Equal Vectors

Exercise Set 8.1

1. If the first quadrant of an xy plane is placed on a hardball field (90 ft. between bases) as in Example 5.3, what are the displacement vectors of:

a.) A single
b.) A double
c.) A triple
d.) A homerun
e.) Advancing from first base to second
f.) Advancing from second base to third
g.) Advancing from third base to home

2. Figure A shows the map of a city. Each line represents a street and each square represents a city block.

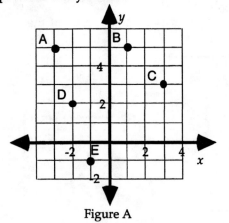

Figure A

Describe the displacement vectors, with units in blocks, to get from:

a.) A to B   e.) C to A   i.) D to C
b.) A to C   f.) C to B   j.) E to B
c.) B to E   g.) C to E   k.) E to C
d.) B to D   h.) D to A   l.) E to E

3. A pizzeria is located at point P on the map below:

H St.	G St.	F St.	E St.	D St.	C St.	B St.	A St.

1st Ave.
2nd Ave.
3rd Ave.
4th Ave.
5th Ave.
6th Ave.
7th Ave.
9th Ave.

P

Figure B

Over the course of an evening, an employee makes a number of deliveries.

The routes the driver takes to make deliveries can be described by the following displacement vectors with units in city blocks.

c.) $3i + 3j$ g.) $2i + 4j$
d.) $-2i - 2j$ h.) $-1i + 0j$

From the vectors determine the locations of the intersections where the deliveries were made.

a.) $-3i + 4j$ e.) $-4i + 2j$
b.) $2i - 2j$ f.) $0i + 2j$

8.2 ADDITION OF VECTORS

In addition to displacement vectors, vectors are used to indicate a number of other things such as velocity and force. One of the main purposes for breaking things such as force into i, j, and k components is that it allows us to determine the result when a combinations of forces act on an object. When a number of forces act on an object and are broken down into i, j, and k vectors, the resulting net force is simply the sum of these vectors. The following discussion provides some examples of how to add velocity and force vectors.

8.2.1 Velocity Vectors

In the chapter on Speed, we looked at how the windspeed of the air will affect the groundspeed of an airplane. In these problems, the wind always traveled in the same direction as the plane. Of course it is unlikely that the wind would have the same direction as the plane. The wind will usually travel at some angle with the direction of the plane. When this happens, the resulting groundspeed is determined using the vectors of the airspeed and windspeed. The two *velocity* vectors would need to be added. When two or more vectors are added together, the combination forms a new vector called the *resultant*. The following example illustrates of how vectors are added.

Example 8.: Al crosses a river rowing at 12 ft./sec. The flow of the river is 5 ft./sec.
a.) What is the resultant velocity of the boat? In other words, how does someone standing on the shore motion of the boat?
b.) perceive the What is the magnitude of the resultant?

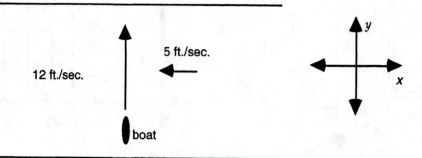

Figure 8.4 Rowing and River Velocity Vectors

<u>Solution:</u> a.)The motion of the boat results from two movements, Al's rowing and the flow of the river. To see how the boat travels, we draw the vectors of these two velocities.

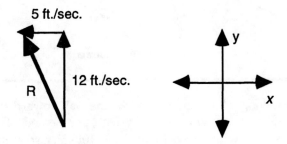

Figure 8.5 Addition of Velocity Vectors

In Figure 8.5, the velocity of the rowing (the y component) is $12j$. The river is flowing in the negative x-direction at 5 ft./sec. The velocity of the river is therefore $-5i$. The resultant vector is found by adding the two velocity vectors head to tail. The resultant's tail is at the tail of the last vector and its head is at the head of the first vector. The resultant is usually shown as bolder than the other vectors.

$$\mathbf{R} = -5i + 12j$$

The resultant velocity vector of Al's boat is $-5i + 12j$

b.) The Pythagorean Theorem can then be used to determine the magnitude of the resultant velocity. Since we know that the magnitude of the speed is positive, the negative root of R is discounted.

$$
\begin{aligned}
R^2 &= (-5)^2 + 12^2 \\
R &= \pm\sqrt{25 + 144} \\
&= \sqrt{169} \\
&= 13
\end{aligned}
$$

The magnitude of the resultant vector is 13 ft./sec.

The same resultant vector is created regardless of the order in which they're added. The addition of vectors is therefore said to be commutative.

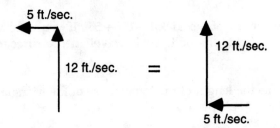

Figure 8.6 Vector Addition is Commutative

The subtraction of a vector is the same thing as adding a vector of equal magnitude in the opposite direction.

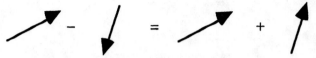

Figure 8.7 Vector Subtraction

Example 8.2: As a field goal kicker, John kicks the ball toward the goal posts with a velocity of $V_k = -5i + 60j$, with units in feet per second. The playing field is experiencing a cross wind with a velocity of $V_w = -10i - 5j$, also with units of ft./sec.

a.) What is the resultant velocity vector of the ball while it travels through the air?
b.) What is the actual speed of the ball?

Solution: We can visualize the problem by graphing the vectors head to tail.

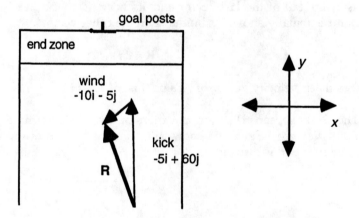

Figure 8.8 Velocity Vectors of Example 8.2

a) To determine the resultant velocity vector, we simply add the two velocity vectors.

$$
\begin{array}{r}
-5i + 60j \\
+ \quad -10i - 5j \\
\hline
= \quad -15i + 55j
\end{array}
$$

The resulting velocity vector of the kick is $-15i + 55j$. If you were watching the game from a blimp above the field, the ball would travel in the direction of the resultant vector.

b.) The speed represents the length of the hypotenuse of the triangle determined by the resultant vector. So the actual speed is:

$$\sqrt{(-15)^2 + (55)^2}$$

$$= \sqrt{3250}$$

$$\approx 57 \text{ ft.}/\text{sec.}$$

The actual speed of the football is 57 ft./sec.

8.2.2 Force Vectors

Vectors are also used to determine the result of several forces acting on an object. Force vectors are added in the same that velocity vectors are. When vectors describe forces, the units are in pounds or the metric unit of force, the *Newton* (N).

Example 8.3: During a running play in a football game, an offensive guard and tackle double-team a defensive guard, as shown in Figure 8.9, to create a hole for the running back. As the block is made, all three players effectively move as one unit. The players move as a result of the combination of forces that each applies. The play is viewed from above the field. The force vectors of the players are as follows:

defensive guard, DG: -220*j*

offensive guard, OG: 5*i* + 200*j*

offensive tackle, OT: -40*i* +140*j*   where the units of the vectors are in pounds.

(The forces are created by the players' legs. The magnitude of the leg forces is like the leg force needed to move the same amount of weight in a squat lift.)

Determine the vector of the resultant force and its magnitude.

Solution: As with the velocity vectors, the resultant of the addition of force vectors can be determined graphically by adding the force vectors head to tail. The three vectors are graphed as shown in Figure 8.9.

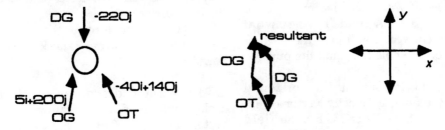

Figure 8.9 Force Vectors of Example 8.3

The resultant vector is determined by adding the three force vectors:

DG   -220*j*

OG $5i + 200j$
OT $\underline{-40i + 140j}$
                                            $-35i + 120j$

The resultant vector tells us that the net force of the blocks was a little to the left ($-35i$) but mostly forward ($120j$).

The magnitude of the resultant is:

$$= \sqrt{35^2 + 120^2}$$

$$= \sqrt{15625}$$

$$= 125 \text{ lbs.}$$

The magnitude tells us how much the defensive guard is pushed.

The defensive guard is pushed back and to the left by a force of 125 lb. by the two offensive linemen.

Exercise Set 8.2

1. During a particular hole on a golf course, a golfer's strokes may be described with the following displacement vectors with units in yards:

Stroke	Displacement Vector
1	$-23i + 162j$
2	$-12i + 34j$
3	$6i + 16j$
4	$1i - 2j$
5	$-1i + 1j$

Table A

a.) What displacement vector would have given the golfer a hole in one?

b.) The golfer's partner lands on the green after a tee-off with a displacement vector of $-28i + 209j$. A putt having what displacement vector will give the partner a birdie on the hole (i.e., sink the putt)?

2. A softball is hit with a horizontal trajectory having a velocity vector of $62i + 28j$. At the time the ball is hit, the field is experiencing a cross wind with a velocity vector of $3i - 4j$, both with units in ft./sec.

a.) What is the resulting velocity vector of the ball's travel?

b.) What is the magnitude of the ball's speed?

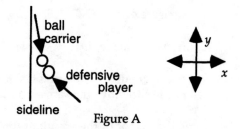

Figure A

3. During a tackle at the sideline, two players hit each other and move as a unit. The force generated by the defensive player is $-34i + 22j$. The force generated by the ball carrier is $8i - 46j$.

a.) What is the net force of the tackle?

b.) Is the ball carrier knocked out of bounds?

4. To arrive at its destination on schedule, a plane will need to maintain a velocity vector of $-342i + 235j$ (groundspeed) with units in miles per hour. During the flight, the plane is expected to experience a cross wind of $27i - 16j$ (airspeed). At what velocity vector should the pilot fly the plane?

5. A punt is kicked on a horizontal trajectory with a velocity vector of 38*i* - 24*j*. At the time the ball is kicked, the field is experiencing a cross wind with a velocity vector of 6*i* - 15*j*.
a.) What is the resulting velocity vector of the ball's travel?
b.) What is the magnitude of the ball's speed?

6. During a goal line stand in a football game, three offensive players are locked with three defensive players and are effectively moving as a unit. The force vector generated by each player is shown in Table B with units in pounds.

Player	Force Vector
Fullback	-20*i* - 180*j*
Offensive guard	25*i* - 200*j*
Offensive tackle	-10*i* - 200*j*
Defensive guard	15*i* + 180*j*
Defensive tackle	15*i* + 190*j*
Linebacker	5*i* + 180*j*

Table B

To help stop the progress, another defensive player joins in. What is the minimum *j*-component of the safety's force vector required to stop the offense?

Figure B

7. The body of a water-skier is acted on by a number of different forces. The tow rope pulls him with a force vector of 75*i* - 12*j* - 0*k*. Gravity acts on his body giving his weight a vector of 0*i* - 0*j* - 160*k*. The water creates a reaction with a vector of -40*i* - 10*j* - 160*k*. Wind resistance creates a force with a vector of -20*i* - 0*j* -0*k*. What is the net force vector on the water skier?

8. A backboard assembly in its retracted position is supported by four poles and two cables (which connect together) as shown in Figure C.

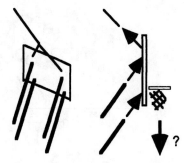

Figure C

To help support the weight of the backboard, each pole generates a force on the backboard with a vector of 50*i* - 50*j* - 50*k* with units in pounds. To support the backboard, the two cables generate forces of -100*i* – 100*j* - 100*k* and -100*i* - 100*j* -100*k*. For the system to balance, all of the forces must sum to zero in every direction. The only other force on the backboard is gravity. Determine the weight of the backboard assembly.

8.3 ROTATION VECTORS

As stated at the beginning of the chapter, all motion can be divided into linear components and rotation. Rotation can also be defined using *ijk* unit vectors. The unit rotation vectors describe the axis of rotation and its magnitude. Since an object can rotate in two different directions on the same axis, defining the axis alone can't describe the rotation. Rotation in one direction is considered positive while rotation in the other direction is considered negative. To determine which is which, scientists and engineers use a principal called the *right hand rule*. If the curved fingers of the right hand point in the direction of the rotation, the extended thumb will point in the positive direction of the axis.

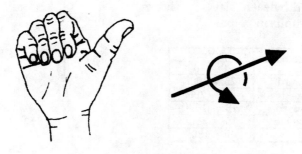

Figure 8.10 The Right Hand Rule

The magnitude of the rotation vector is determined by the speed at which it's spinning or its *angular velocity* . The magnitude of the angular velocity is given by the number of rotations per unit time. The units of *revolutions per minute* or *rpm's* used with machinery is a unit of angular velocity. Angular velocity is usually denoted with the Greek letter ω (pronounced "omega") which is the lower case letter for "o".

The different types of pitches thrown in baseball help describe how rotation is defined. For this description, we will assume that all of the pitches have an angular velocity of a revolutions per second. All of the pitches will be defined as being thrown by a right handed pitcher away from the pitcher.

If we put the center of the ball at the origin of the *xyz* graph as shown in Figure 8.11, we can define its rotation by the *ijk* unit vectors. The rotations shown in Figure 8.12 are approximate. The actual rotation axis of a pitched ball will probably never be perfectly horizontal or vertical.

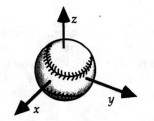

Fig. 8.11 Orientation of Axes

The typical fastball is thrown with bottom spin and doesn't curve to the left or right. It spins on a horizontal axis. If you curve the fingers of your right hand in the direction of the spin, your extended thumb points in the positive direction of the y-axis. Its rotation vector is therefore $+aj$. A drop ball is given top spin. It spins on the same axis as the fastball but in the opposite direction. Its rotation vector is therefore $-aj$. A slider is meant to curve to the left. Its axis is vertical. It has a rotational vector of $+ak$. [1] The screwball is meant to curve to the right, also spinning on a vertical axis but in the opposite direction of a slider. Its rotation vector is $-ak$. The curve ball is a combination of a drop ball and a slider (preferably more drop than slider). Its axis is tilted creating a rotation vector of about $-bj + ck$, where $b^2 + c^2 = a^2$. What about the knuckle ball? If thrown right, the knuckle ball isn't given any spin. Its rotation vector is $0i + 0j + 0k$.

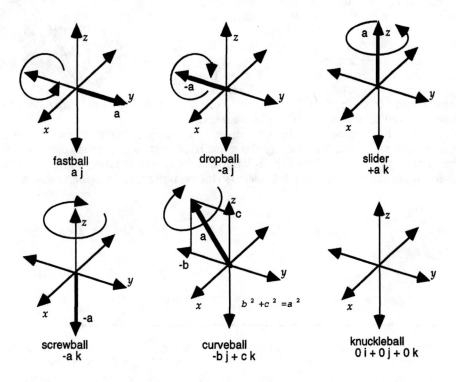

Figure 8.12 Rotations of Breaking Pitches

---

[1] Some players define a slider as also having a drop component like a curveball.

8.3.1 Radians

In the mechanical world, a need often arises to determine the linear displacement which results from rotation. For example, it might be necessary to determine how far a wheel with a 12-in. radius would move a bicycle after the wheel has rotated 150°. We know that if the wheel rotated 360°, the displacement would be equal to the circumference of the wheel which is equal to $2\pi r$. To rotate the wheel 150° would move the bike a fraction of the circumference equal to:

$$\frac{150°}{360°}(2\pi r) = \frac{5}{12}(2\pi)12$$

$$\approx 31.4 \text{ in.}$$

In Figure 8.13, the wheel has rotated to an angle θ, or theta, which is the Greek letter for the "th" sound. A system of measuring angles (like the degree system), has been developed to allow us to calculate the linear distances directly from the rotation and the length of the radius. The unit of this system is the *radian*. Rather than calculating fractions of the degree system, the radian is simply defined as the angle necessary to rotate a point on a circle a distance equal to the radius. An angle of one radian is equal to approximately 57.3 degrees.

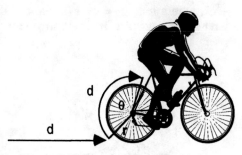

Figure 8.13 Bicycle Wheel Rotation

Every circle has 2π radians in it. Since it is so easy to convert between linear displacement and rotation when radians are used, rotation vectors usually have their units in radians. Eq. 8-1 shows the relationship between distance, radius, and angles in radians. A similar equation, Eq. 8-2 relates velocity, radius, and angular velocity.

Figure 8.14 The Radian

1 radian ≈ 57.3°

$$d = r\theta \qquad\qquad (8-1)$$

$$v = r\omega \qquad\qquad (8-2)$$

Example 8.4: Jill works her triceps muscles on a machine as shown in Figure 8.15. As she pulls the bar down, the pulleys rotate in the direction of the cable. If the pulley wheels have a radius of 6 inches (0.5 ft.), how many radians will they rotate when she moves the bar 1 ft.?

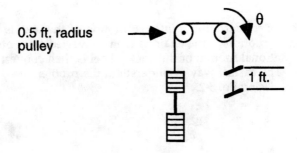

0.5 ft. radius
pulley

θ

1 ft.

Figure 8.15 Pulley Rotation

Solution: In this problem, we need to solve Eq. 8-1 for θ. We do so by dividing both sides by r.

$$d = r\theta$$

$$\theta = \frac{d}{r}$$

We then substitute the radius of the wheels and the distance the bar moved into the equation.

$$\theta = \frac{d}{r} = \frac{1 \text{ ft.}}{0.5 \text{ ft.}} = 2 \text{ radians}$$

When the bar is lowered 1 ft., the wheels rotate 2 radians.

Example 8.5: A cyclist is pedaling at a rate which rotates her front gears at an angular velocity of ω_f. The front gear with the chain on it has a radius of r_f. The gears, of course, pull a chain which travels at a velocity of v_c. The chain pulls the gears on the rear wheel which rotates the wheel at an angular velocity of ω_w. The gear at the rear wheel has a radius of r_r. The radius of the tire is r_t. As a result of the rear wheel rotating, the bicycle travels at a velocity of v_b. Given the configuration, develop an equation which gives the value of v_b in terms of ω_f, r_f, r_r, and r_t.

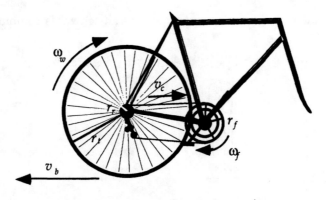

Figure 8.16 Bicycle Rotations and Translations

<u>Solution:</u> The bicycle converts the rotational motion of the pedals to a linear motion of the chain. The linear motion of the chain is then converted to a rotational motion of the back wheel. The rotational motion of the back wheel is then converted to a linear motion of the whole bike. The easiest way to understand the problem is to develop an equation for each conversion. Using Eq. 8-2:

At the pedal:

$$v_c = r_f \omega_f$$

v_c = velocity of chain

ω_f = angular velocity of front gear

r_f = radius of front gear

At the gears on the rear wheel:

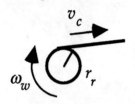

$$\omega_w = \frac{v_c}{r_r}$$

v_c = velocity of chain

ω_w = angular velocity of rear wheel

r_r = radius of gear on rear wheel

At the pavement:

The angular velocity of the tire is the same as the angular velocity of the rear gears.

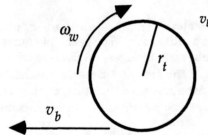

$$v_b = r_t \omega_w$$

v_b = velocity of bike

ω_w = angular velocity of rear wheel

r_t = radius of tire

To get v_b as a function of ω_f, r_f, r_t, and r_r, we work backwards combining the equations we just developed:

$$v_b = r_t \omega_w \qquad \omega_w = \frac{v_c}{r_r} \qquad v_c = r_f \omega_f$$

$$= r_t \left(\frac{v_c}{r_r} \right)$$

$$= r_t \left(\frac{r_f \omega_f}{r_r} \right)$$

$$= \frac{r_t r_f \omega_f}{r_r}$$

The velocity of the bike is given by the equation: $v_b = \dfrac{r_t r_f \omega_f}{r_r}$

Exercise Set 8.3

1. A certain type of pitching machine uses two large spinning rubber discs to propel the ball toward the batter. The discs are 1 ft. in diameter. The angular velocity of the discs is 88 radians/sec. If the ball continues at the speed of the outside edge of the discs, what is the velocity of the ball in ft./sec.? In mi./hr.?

2. An auto shop which specializes in changing oil dispenses the oil from hoses which are coiled on spools overhead. A spring on the spool recoils the tube when it is not in use. The farthest position the hose will need to reach in the shop is a point 6 ft. below the spool, 20 ft. in front of it, and 10 ft. to the right. The spool has a radius of 1 ft. How many radians must the spool be able to recoil to serve the entire shop?

3. A certain type of appliance is assembled along an assembly line in a factory. A conveyor belt moves the appliances to different workers as they add parts. The conveyor belt is driven by rollers with 3-in. diameters. If the conveyor belt moves at 2 in./sec., what is the angular velocity of the rollers in radians per second? In revolutions per minute?

4. Example 8.5 derived the following equation:

$$v_b = \frac{r_t r_f \omega_f}{r_r}$$

where v_b = the velocity of the bike

r_t = the radius of the tires
r_f = the radius of the front gears
ω_f = the angular velocity of the front
    gears
r_r = the radius of the rear gears

If r_t, = 1.2 ft., r_f, =0.3 ft., and r_r = 0.1 ft., determine how fast the cyclist must pedal (in radians/sec.) to move the bike 10 mph, 30 mph, and 45 mph.

5. A company is designing a jogging machine. A treadmill will rap around a set of rollers. The rollers have a radius of 0.1 ft. A mechanism is placed on one of the rollers which measures its angular velocity in radians per second. The jogging machine will have an electronic display which shows the speed of the jogger in miles per hour. To program the electronic display, the manufacturer needs a multiplier, m, to convert the angular velocity of the rollers into miles per hour.

$v = m\omega$ v = velocity in mph
          ω = angular velocity in
          rad./sec.

Determine the value of the multiplier, m.

6. School children play a jump rope game called "double dutch" which involves two ropes rotating at the same speed but in opposite directions. If the angular velocity vector of one of the ropes is $3i - 4j + 0 k$, what is the angular velocity vector of the other rope?

7. Pilots and aircraft designers have given special names for each direction that a plane may rotate. As shown in Figure A, when a plane rotates about an axis along its fuselage (the x-axis), the rotation is known as *roll*. When it rotates about an axis along its wings (the y-axis), the rotation is known as *pitch*. When it rotates along a vertical axis (the z-axis), the rotation is known as *yaw*.

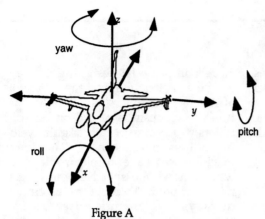

A plane rotates to an angle, θ, about one of these axes. Determine the rotation vector for each of the following rotations:

    a.) yaw to the right
    b.) yaw to the left
    c.) pitch up
    d.) pitch down
    e.) roll to the right
    f.) roll to the left

Figure A

8.4 MULTIPLICATION OF VECTORS

Vectors can be multiplied in two ways. When a vector is multiplied in a way which only affects its magnitude, the result is a *scalar product*. A scalar product can also be described as the product of a vector and a scalar. When two vectors are multiplied, a *cross product* is produced.

8.4.1 Scalar Product

A vector can be multiplied in a way which affects its magnitude alone. This results in what is known as a scalar or *dot* product.

Example 8.6: The pin of a squat machine is set at 20 lb. This creates a force vector (neglecting the weight of bars) of -20 k with units in pounds. What is the force vector if the weight is tripled?

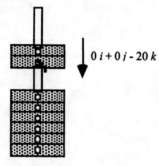

$$0\,i + 0\,j - 20\,k$$

Figure 8.17 Tripling Weight

Solution: If the weight is tripled, the force vector is multiplied by 3. The number three is a scalar number; it has no direction. The answer is therefore a scalar product. To triple this force would be to set the pin at 60 lb.

$$3 \cdot (0\,i + 0\,j - 20\,k) = 0\,i + 0\,j - 60\,k$$

The force vector created by tripling the weight is $0\,i + 0\,j - 60k$.

8.4.2 Cross Products

When two vectors are multiplied, the result is a cross or *vector* product. Graphically, when two vectors are multiplied, the result is a vector perpendicular to the original two vectors. The determination of the perpendicular vector has significance for types of motion that are common in sports and machinery. We're often interested in perpendicular vectors in the conversion from angular velocity to velocity and from force to torque.

In Chapter 3, we saw that the moment created by the weights of a bench machine equals the force of the weights times the length of the moment arm. But we could only solve problems when we knew the perpendicular distance between the hinge and the line of force. If we define the bar in terms of a displacement vector and the weight in terms of a force vector, we can find the resulting moment by the use of vector products or cross products. With vectors, we can determine the torque when the force is acting at any obscure angle and the moment arm is at an obscure angle. The letter tau, τ, the lower case Greek letter for "t" is usually used to denote torque. The vector formula to calculate torque, τ, is:

$$\tau = r \times F \qquad\qquad (8\text{-}3)$$

The standard equation used to find the cross product of two vectors, A and B, is as follows:

$$A = a_1 i + a_2 j + a_3 k$$
$$B = b_1 i + b_2 j + b_3 k$$

$$(8\text{-}4)$$

$$\begin{vmatrix} i & j & k \\ a_1 & a_2 & a_3 \\ b_1 & b_2 & b_3 \end{vmatrix} = i \begin{vmatrix} a_2 & a_3 \\ b_2 & b_3 \end{vmatrix} - j \begin{vmatrix} a_1 & a_3 \\ b_1 & b_3 \end{vmatrix} + k \begin{vmatrix} a_1 & a_2 \\ b_1 & b_2 \end{vmatrix}$$

The determinants are calculated with the same procedure used with Cramer's Rule. Since the right hand rule produces different results depending on the order in which you cross the vectors, the multiplication of vectors is not commutative. You must always be sure that when you use Eq. 8-4 to determine the torque vector, the determinant has the radius vector in the middle row and the force vector in the bottom row.

Example 8.8: An athlete is developing his pectoral muscles on a cable crossover machine as shown in Figure 8.18.

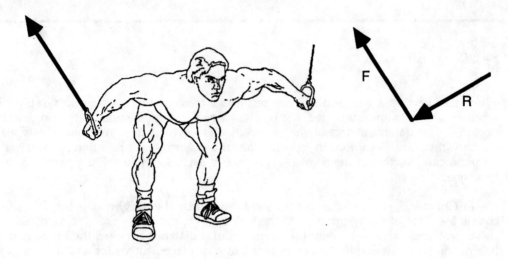

Figure 8.18 Cross Product of Cable Crossover

At the position shown, the cable creates a force with a vector of:

$$\mathbf{F} = 80i - 40j + 80k \qquad \text{where the units are in pounds}$$

The athlete's arm is extended to a displacement or *radius* vector of

$$\mathbf{R} = 1.25i + 0.5j - 0.5k \qquad \text{where the units are in feet}$$

What is the vector which represents the torque on his shoulder?

Solution: Recall that torque is the product of force times radius. The first step in solving a cross product problem is to use the right hand rule to determine the general direction of the cross product. The vectors are placed tail to tail.

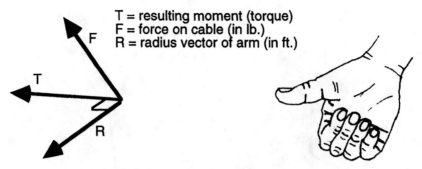

T = resulting moment (torque)
F = force on cable (in lb.)
R = radius vector of arm (in ft.)

Figure 8.19 Use of Right Hand Rule

Using the right hand rule, if your fingers start at the position of the radius vector and cross to the position of the force vector, your thumb will point back and to the left. The torque vector is therefore as shown in Figure 8.19. From the figure, you can see that the torque vector is at the axis about which his shoulder would rotate. The determinant of the problem is:

$$T = R \times F =$$

$$\begin{vmatrix} i & j & k \\ 1.25 & 0.5 & 0.5 \\ 80 & -40 & 80 \end{vmatrix} = i \begin{vmatrix} 0.5 & -0.5 \\ -40 & 80 \end{vmatrix} - j \begin{vmatrix} 1.25 & -0.5 \\ 80 & 80 \end{vmatrix} + k \begin{vmatrix} 1.25 & 0.5 \\ 80 & -40 \end{vmatrix}$$

$$= (40 - 20)i - (100 - (-40))j + (-50 - 40)k$$

$$= 20i - 140j - 90k$$

The resulting torque vector on the shoulder is $T = 20i - 140j - 90k$.

A similar cross product relationship exists between velocity, angular velocity, and the radius arm. The vector formula of the relationship is:

$$v = \omega \times r \qquad (8\text{-}5)$$

This relationship is also not commutative. The right hand rule must be used to determine the positive direction of the velocity vector.

Example 8.8: A tennis player serves the ball. At the time he hits the ball, his arm and racket create a radius vector of:

$$r = 1i + 1j + 3k \qquad \text{where the units are in feet.}$$

The radius vector extends to the point on the racket where the ball is hit. His arm and racket rotate about his shoulder with a vector of:

$$\omega = 40i - 5j + 5k \qquad \text{where the units are in radians per second}$$

If the ball continues at the speed of the racket, what is the velocity vector of the ball at the time it is hit?

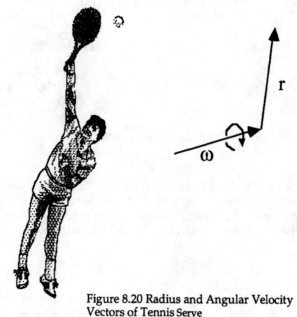

Figure 8.20 Radius and Angular Velocity
Vectors of Tennis Serve

<u>Solution:</u> Again, we first use the right hand rule to determine the general direction of the cross product. We put the radius and angular velocity vectors tail to tail. If your fingers cross from the ω vector to the **r** vector, your thumb sticks out forward and to the right. The velocity vector is therefore as shown in Figure 8.21. When you look at the figure of the tennis player serving, the direction of the velocity vector makes sense. This is the direction the ball will travel.

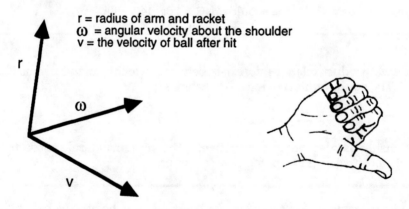

Figure 8.21 Right Hand Rule in Determining Velocity Vectors

$\mathbf{v} = \omega \times \mathbf{r} =$

$$\begin{vmatrix} i & j & k \\ -40 & -5 & 5 \\ 1 & 1 & 3 \end{vmatrix} = i \begin{vmatrix} -5 & 5 \\ 1 & 3 \end{vmatrix} - j \begin{vmatrix} -40 & 5 \\ 1 & 3 \end{vmatrix} + k \begin{vmatrix} -40 & -5 \\ 1 & 1 \end{vmatrix}$$

$$= (-15 - 5)i - (-120 - 5)j + (-40 - (-5))k$$

$$= -20i + 125j - 35k$$

The initial velocity vector of the tennis ball after it is hit is $\mathbf{v} = -20i + 125j - 35k$ where the units are in feet per second. The actual magnitude of the velocity is:

$$v = \sqrt{(-20)^2 + (125)^2 + (-35)^2}$$

$$\approx 131 \text{ ft./sec.}$$

In the next chapter, we're going to look at how the velocity vector can be use to determine the path of the ball.

It is important to note some of the simplifications of these problems. The full motion of doing a cable crossover or serving a tennis ball is much more complex than the problems shown here. The tennis player would neither keep his arm perfectly straight nor rotate it about his shoulder at a constant speed through the entire serve. The calculations shown here reflect only the motion taking place a particular instant.

Exercise Set 8.4

1. An inclined leg press is loaded with 100 lb. Since the press is inclined the weights create a force vector of $0i - 80j - 60k$. What is the resulting force vector if the weight is reduced by one half? (Neglect the weight of the equipment.)

2. A speed skier is skiing down a hill with a constant slope giving her a velocity vector of $0i + 24j - 18k$ with units in mph. By the time she is approaching the bottom, she has quadrupled her speed.
a.) What is her velocity vector at this point?
b.) What is her speed?

3. A golfer tees off with her arms and shoulders rotating with an angular velocity vector of $0i - 26j + 8.5k$ with units in radians per second. Her arms and club extend to a radius vector of $3i - 20j - 60k$ with units in inches.

a.) What is the resulting velocity vector of the ball (in ft./sec.) after being hit?
b.) What is the speed of the ball in ft./sec.?

4. At a track meet, a hammer thrower releases the hammer with his arms and body rotating with an angular velocity vector of $0i - 40j + 11k$ in radians per second. His arms and the hammer have a radius vector of $4i - 73j + 27k$., with units in inches. What is the vector which describes the velocity of the ball as it is released?

5. On a backhand return, a tennis player's arm and racket rotate about her shoulder with an angular velocity having an angular velocity vector of $0i - 1.3j + 15k$, with units in radians per second. The point at which the ball hits the racket creates a radius vector from the shoulder of $1i - 41j + 4.2k$, with units in inches. What is the initial velocity vector of the tennis ball after being hit?

6. A school is installing a new backboard and pole on one of its outdoor basketball courts. Unfortunately, the pole has been set in its hole in the backboard at a wrong angle with the baseline and must be rotated. To rotate the pole, a rope is tied around the rim. The location where the rope is tied creates a moment arm on the pole with a displacement vector of $5i - 1j + 10k$ with units in ft. By pulling on the rope, a force vector of $80i - 15j + 60k$, with units in lb., is created. What is the resulting torque vector on the pole?

Determine the following vector products:

7. $12 \cdot 5i - 4j$

8. $0.25 \cdot 2i - 16j + 1k$

9. $3i - 1j \times 4i + 8j$

10. $\dfrac{1}{3} \cdot 24i + 36j - 6k$

11. $6i + 0j + 3k \times 0i - 2j - 4k$

12. $0.5i - 1j + 2.2k \times 3i - 0.2j + 1.5k$

13. $0i - 0j - 3k \times i - 0j + 0k$

14. $3i - 5j + 1k \times 8i - 2j + 10k$

Chapter 9

PARABOLAS

In Chapter 6, we studied the motion of objects or people traveling at constant speeds. In Chapter 8, we learned how forces acting in different directions create motion in a particular path. By building on these two concepts, we can understand one of the most common phenomena in sports - how objects travel in the air.

Footballs, basketballs, baseballs, soccer balls, volleyballs, tennis balls, golf balls, shot puts, high jumpers, long jumpers, gymnasts - they all travel through the air in the same kind of path. The path is called a parabolic curve or a *parabola*. Parabolas and *quadratic equations*, the equations that describe their paths, occur in innumerable areas in the professional world.

9.1 PARABOLIC TRAVEL

Figure 9.1 illustrates the characteristics of the parabolic travel of a ball through the air. The figure shows a side view of the parabolic arc, a top view to show travel in only the x-direction, and an end view to show travel in only the y-direction. The figure shows the position of the ball at constant time intervals, for example, every half-second or so. Note that in the x-direction, the travel distance for each time interval remains the same. This is because the object is given an initial velocity in the x-direction when thrown and nothing opposes it in the x-direction (except for air resistance which is ignored in this chapter). In the y-direction, the distance traveled between each time interval decreases on the way up and increases on the way down. The object is given an initial velocity in the y-direction. Gravity acts on the object for its entire path. On the way up, gravity slows down the object in the y-direction until it overcomes the initial upward velocity. At the peak of the curve, gravity has slowed down the velocity in the y-direction to zero. On the way down, gravity accelerates the object to fall faster and faster. The longer gravity has to act on an object, the faster it will travel.

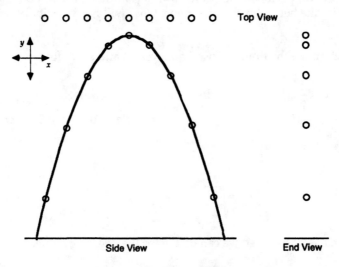

Figure 9.1 Parabolic Travel

Another important aspect of parabolic travel is viewed from the perspective of the ball's conversion between different types of energy. When the ball is first thrown, its speed is at its greatest. For the rest of the way up, gravity will act to slow it down. Since the ball's speed is at its greatest when it's first thrown, it has its greatest *kinetic energy* at this time. As the ball continues to travel upward, its speed decreases. When the ball reaches its peak, its speed, and therefore its kinetic energy, are at their lowest. However, since the ball has reached this height, its *potential* energy is at its greatest. The energy is called potential because can gravity can pull on it to get it moving again. Water behind a dam also has potential energy because it can be run through turbines to generate electricity. As gravity continues to pull on the ball on its way down, it accelerates it increasing its kinetic energy. Just before it hits the ground, the ball again reaches its maximum velocity and therefore its maximum kinetic energy. At this point, the speed of the ball will equal its speed when it was first thrown. However, since the ball has lost its height on the way down, it has also lost its potential energy. Along the trip, therefore, the ball undergoes a conversion from kinetic energy to potential energy and from potential energy to kinetic energy.

9.1.1 Gravitational Acceleration

To better understand how objects travel through the air, it is important to be familiar with the concept of *gravitational acceleration*. If you ignore factors such as aerodynamic qualities of an object (like a Frisbee) or the particular elevation of the area where it's thrown, the rate of gravitational acceleration on earth is relatively constant. The rate at which the earth accelerates objects is known as the *gravitational constant*. In the US system of measurements, the gravitational constant, g, is 32.2 ft/sec^2. In metric units, g is 9.81 meters/sec.^{2}. Here are two physical formulas involving gravitational acceleration:

$$d = \frac{1}{2}at^2 \quad \text{(9-1)} \quad \text{where} \quad d = \text{the vertical distance} \quad a = \text{acceleration}$$

$$v = at \quad \text{(9-2)} \quad t = \text{time} \quad v = \text{velocity}$$

Example 9.1: Brenda performs her dive from a platform 50 feet above the water level. How long will it take her to hit the water and what will her velocity be when she hits the water?

Solution: To determine the time, we use Equation 9-1 and solve for t.

$$d = \frac{1}{2}at^2 \qquad d = \text{distance} \qquad t = \text{time}$$

$$2d = at^2 \qquad a = \text{acceleration} = g = 32.2 \text{ ft/sec.}^2$$

$$t^2 = \frac{2d}{a}$$

$$t = \pm\sqrt{\frac{2d}{a}}$$

We know the time is positive, so we can disregard the negative value.

$$t = \sqrt{\frac{2(50)}{32.2}} = 1.76 \text{ sec.}$$

It takes 1.76 seconds for Brenda to hit the water.

To determine her velocity when she hits the water, we put 1.76 seconds into Eq. 9-2.

$$v = at = gt = (32.2 \text{ ft.}/\text{sec.}^2)(1.76 \text{ sec.}) = 56.7 \text{ ft.}/\text{sec}$$

When Brenda hits the water, she is traveling at 56.7 feet per second.

Note that Brenda's velocity increases with time. The fact that she's traveling faster and faster is why g is known as the *acceleration* of gravity.

9.1.2 Parabolic Trajectories

By combining the acceleration of gravity with the velocity of the object when it is first thrown, formulas can be derived which show the path of the object as it moves through the air. The following formulas give the position and velocities of an object as a function of time:

$$x = v_{0x}t \qquad (9\text{-}3) \qquad\qquad v_x = v_{0x} \qquad (9\text{-}5)$$

$$y = -\frac{gt^2}{2} + v_{0y}t \qquad (9\text{-}4) \qquad\qquad v_y = v_{0y} - gt \qquad (9\text{-}6)$$

where v_{0x} = the initial velocity of the object in the x-direction
v_{0y} = the initial velocity of the object in the y-direction
g = the gravitational constant
t = the time elapsed since the object was thrown

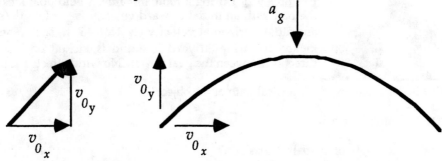

Figure 9.2 Vectors inolved in Parabolic Travel

The 0's in the terms indicate that the terms represent conditions at time equals zero. The v_{0x} and v_{0y} terms represent the velocity vectors of the initial trajectory of the object. In the xy plane, the initial velocity vector could be written as $v_{0x}i + v_{0y}j$.

The velocity in the x-direction is assumed to remain constant. Note that Eq. 9-3 is virtually identical to the $d = rt$ equation discussed in the Speed chapter.

The velocity in the y-direction, v_{0y}, changes because of the pull of gravity. In the equation $y = -\dfrac{gt^2}{2} + v_{0y}t$, the term "$v_{0y}t$" represents the height the object would be at without gravity. The "$-\dfrac{1}{2}gt^2$" term represents the distance gravity has pulled it back. If the travel of an object is plotted on an xy plane, its position is given by the equation:

$$y = -\frac{g}{2v^2_{0x}}x^2 + \left(\frac{v_{0y}}{v_{0x}}\right)x \qquad (9\text{-}7)$$

Other equations include:

$$H = \frac{v^2_{0y}}{2g} \qquad \text{where } H = \text{the maximum height reached by the object} \qquad (9\text{-}8)$$

$$R = \frac{2v_{0x}v_{0y}}{g} \qquad \text{where } R = \text{the range of travel} \qquad (9\text{-}9)$$

Equations 9-8 and 9-9 assume that the object begins its travel at the ground level and that the land in the area is flat. To determine heights achieved by objects starting their travel above the ground (like a baseball being hit), you need to add the height of the point where the travel began.

For flexible objects like a human body or for spinning objects, it is the center of gravity of the object which stays on a constant parabolic path.

Example 9.2: During a football game, a field goal kicker kicks the ball with an initial upward velocity (v_{0y}) of 46 ft./sec. and an initial horizontal velocity (v_{0x}) of 46 ft./sec. The ball is kicked from the 30 yard line and is kicked on-line (in the direction between the goal posts). No wind affects the ball.

a.) What is the equation of the vertical path of the object?

b.) Plot the path of the ball.

c.) What is the height of the ball at 1 second?

d.) What is the maximum height (H) reached by the ball?

e.) If the ball is not stopped before it hits the ground, what is the horizontal distance (R) that the ball travels?

f) Is the field goal good?

Solution: If the ball is kicked from the 30-yard line, it is 40 yards from the goal posts. (The end zone is 10 yards deep.) Since the gravitational constant is given in units of ft./sec.², we need to convert the distances from yards to feet. A distance of 40 yards is equivalent to 120 feet.

a.) The path of the object can be determined from Equation 9-7,

$$y = -\frac{g}{2v^2_{0x}}x^2 + \left(\frac{v_{0y}}{v_{0x}}\right)x$$

$$= -\frac{32.2 \text{ ft.}/\text{s}^2}{2(46 \text{ ft.}/\text{s})^2}x^2 + \left(\frac{46 \text{ ft.}/\text{s}}{46 \text{ ft.}/\text{s}}\right)x \qquad \text{with } x \text{ and } y \text{ in ft.}$$

$$= -0.00761x^2 + x$$

The vertical path of the ball is given by the equation: $y = -0.00761x^2 + x$

b) By calculating the value of y every 10 ft. or so, the following data are obtained:

x	10	20	30	40	50	60	70	80	90	100	110	120	130
y	9.24	16.96	23.15	27.83	30.98	32.61	32.72	31.30	28.37	23.49	17.93	10.43	1.41

If the data are plotted, the following graph is obtained:

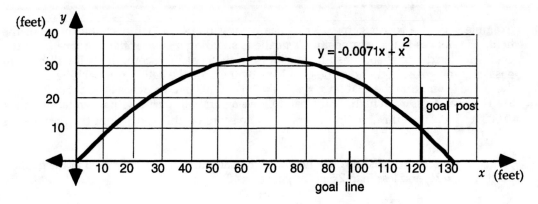

Figure 9.3 Path of Fieldgoal Kick

c.) The height of the ball at a certain time is given by Eq. 9-4.

$$y = -\frac{gt^2}{2} + v_{0y}t$$

$$\text{at } t = 1 \quad y = -\frac{32.2 \text{ ft.}/\text{s}^2}{2}(1 \text{ sec.})^2 + (46 \text{ ft.}/\text{s})(1 \text{ sec.})$$

$$= -16.1 + 46 = 29.9 \text{ ft.}$$

d.) The maximum height is given by Eq. 9-8.

$$H = \frac{v^2_{0y}}{2g} = \frac{(46 \text{ ft.}/\text{s})^2}{2(32.2 \text{ ft.}/\text{s}^2)} = 32.9 \text{ ft.}$$

The maximum height of the ball is 32.9 ft.

e.) The total distance is given by the range equation, Eq. 9-9.

$$R = \frac{2v_{0x}v_{0y}}{g} = \frac{2(46)(46)}{32.2}$$

$$= 131.4 \text{ ft.}$$

f.) Since the problem states that the kick is on-line, we only need to know if it clears the 10 ft. high bar ($y \geq 10$) at $x = 120$ ft. From part a.):

$$y = -0.00761x^2 + x$$

$$= -0.00761(120)^2 + 120$$

$$= 10.4$$

$$10.4 > 10$$

The kick is good.

Example 9.3: In a soccer game, Carlos has possession of the ball near the center of the field. He notices that the opposition's goalie is standing a substantial distance from the goal. He decides to shoot when he is 40m from the goal. To score, the ball will have to pass over the goalie and under the goal bar. By the time the ball is approaching the goal, the goalie is 4m from the goal (at $x = 40 - 4 = 36$) and his hand reaches a height of 3m ($y = 3$). The goal bar is 2.5m high. If the ball follows a parabolic curve with an equation of $y = -0.0053x^2 + 0.28x$, with units in meters, does he score on this shot? Plot the path of the ball.

Solution: To score on this shot, the height of the ball must be greater than 3 at the goalie's position ($x = 36$) and less than 2.5 at the goal ($x = 40$).

at $x = 36$ $y = -0.0053x^2 + 0.28x,$

$$= -0.0053(36^2) + 0.28(36)$$

$$= 3.2\text{m} > 3.0 \text{ m}$$

The ball passes over the goalie.

at $x = 40$ $y = -0.0053x^2 + 0.28x,$

$$= -0.0053(40^2) + 0.28(40)$$

$$= 2.7\text{m} < 2.5\text{m}$$

The ball passes over the goal bar. Therefore, the shot is missed.

To plot the path of the ball, the value of y is calculated every 5m or so.

x	5	10	15	20	25	30	35	40	45	50
y	1.27	2.27	3.01	3.48	3.69	3.63	3.31	2.72	1.87	0.75

If we plot the data, the graph looks like this:

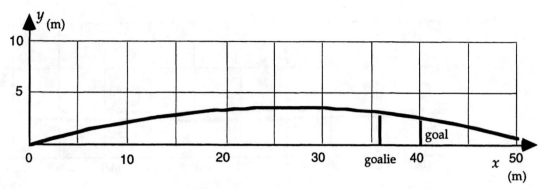

Figure 9.4 Path of Soccer Kick

9.1.3 Vertical Curves

The problems of hitting or missing targets with parabolic curves are very similar to a type of problem faced by civil engineers when designing highways or railroads. Roadways built in hilly terrain require *horizontal curves*, which turn right or left, and *vertical curves*, which curve up or down. Parabolic curves are used as vertical curves on roads over hills and at the bottom of valleys. (Horizontal curves are discussed later in the chapter.)

Figure 9.5 Vertical Curves

When the *xy* coordinate system is used to plot vertical curves, *x* is the distance from the location where the vertical curve starts and *y* is the height or *elevation* of the particular point on the curve.

Example 9.4: A team of civil engineers determines that the most economical route of a planned roadway will have a vertical curve with the equation:

$y = 0.00005x^2 - 0.06x + 100$ where x = the horizontal distance, in feet, and
y = the vertical distance, or *elevation*, in feet.

Another team of engineers is designing an overpass 800 ft. from the point where the vertical curve starts ($x = 800$ ft.). To meet state regulations, the overpass must have an 18 ft. clearance above the highway. The engineers designing the overpass must therefore make sure that the bottom is 18 ft. above the road at that point. To keep costs down, the engineers don't want to make the overpass taller than it needs to be. What is the minimum elevation ($y = ?$) of the bottom of the overpass?

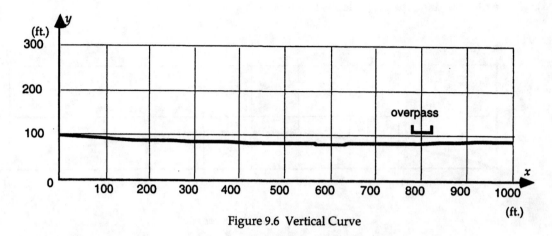

Figure 9.6 Vertical Curve

Solution: To determine the minimum elevation of the bottom of the overpass, we need to find the elevation (y-value) of the roadway at the location below the overpass ($x = 800$).

$$\text{at } x = 800 \text{ ft.} \; y = 0.00005(800 \text{ ft.})^2 - 0.06 \,(800 \text{ ft.}) + 100$$

$$= 84 \text{ ft.}$$

For the bottom of the overpass to be at least 18 ft. above this, the elevation of the bottom of the overpass must equal:

$$84 + 18 = 102 \text{ ft.}$$

The minimum elevation of the bottom of the overpass is 102 ft.

Note - The calculations involved in plotting vertical curves can be made much easier by using computers or programmable calculators. Computers usually recognize exponents with a symbol known as a caret ($\wedge$). For example, with a spreadsheet program, the equation:

$$y = -0.2x^2 + 4 \qquad \text{where } x = \text{the contents of cell A1}$$

would have the spreadsheet equation of:

$$=-0.2(A1\wedge2)+4$$

Most spreadsheet programs can even plot the curve for you.

Exercise Set 9.1

The following problems involve the travel of some objects through the air. Assume that gravity remains constant and that the motion of the object is not affected by air resistance.

1. After clearing the bar, a pole vaulter falls 5.3m to the mat.
a.) How long does it take him to reach the mat?
b.) What is his speed when he hits the mat?

2. A skydiver jumps from a plane at an altitude of 10,000 ft.
a.) If she doesn't open her parachute, what will her altitude be in 5, 10, and 15 seconds?
b.) If she wants to open her parachute at an altitude of 5,000 ft., how many seconds after jumping should she count?
c.) What will her speed be at this point?

3. A pop fly is hit from a height of 3 ft. and given an initial upward velocity, v_{0y}, of 20 ft./sec. Plot the height of the ball every second until in reaches the ground. Plot v_{0y} every second.

4. For flexible objects, the equations for parabolic travel apply to the object's center of gravity. As a high jumper passes over a bar, her back is arched. Her center of gravity will be near her back. The height achieved by her center of gravity will therefore be the same height necessary to clear the bar. To clear the bar, she needs to get her center of gravity to the right height and use the correct technique. The center of gravity of one high jumper's body just before she jumps is located 1m above the ground. Assuming that her technique is perfect, what will her initial upward velocity need to be to clear 2m?

5. A golfer tees-off in an area which is perfectly flat. The initial horizontal velocity is 95 ft./sec. The initial vertical velocity is 90 ft./sec. How long is the drive?

6. While serving, a tennis player hits the ball at a height of 8 ft. straight down the court. The ball is given only an initial horizontal velocity (i.e., no vertical velocity). To clear the net, which is 39 ft. away, the ball must have a minimum height of 3.2 ft. What is the minimum initial velocity required to clear the net?

7. On a pass play, the quarterback throws the ball with a vertical trajectory of $y = -0.0184x^2 + x + 6$ where x is the horizontal distance, in ft. and y is the vertical distance, in ft. A defensive back is on line with the pass and 50 ft. away from the quarterback. How high would the defensive back need to reach to intercept the pass?

8. On a driving range, two golfers notice that their shots landed at the same distance. The shot of the first golfer had an initial horizontal velocity of 99 ft./sec. and an initial vertical velocity of 90 ft./sec. If the second golfer's shot had an initial horizontal velocity of 110 ft./sec., what was its initial vertical velocity?

9. The ceiling in a gym has a height of 40 ft. If an object is thrown from a height of 6 ft., what is the minimum vertical velocity necessary to hit the ceiling?

10. In basketball, the violation of goal tending is called when a player blocks a shot that has started its way downward toward the basket. If a player takes a shot with an initial velocity vector of $18i + 16j$, with units in ft./sec., at what distance from the player will it become illegal to block the shot?

11. A golfer needs to clear some 30-ft. tall trees 150 ft. away. Her shot has an initial velocity vector of $72i + 50j$ with units in ft./sec. and with no wind.
a.) Does the ball clear the trees?
b.) What is the maximum height reached by the ball?
c.) How long is the shot?
d.) Determine the equation that describes the vertical path of the ball.

e.) Plot the path of the ball.

12. A roadway passing over a hill has a vertical curve with an equation of:
$y = -0.00005x^2 + 0.03x + 50$.
a.) Plot the curve
b.) The county would like to put a pedestrian overpass across the road at a location 600 ft. from where the vertical curve starts ($x = 600$). To meet design codes the bottom of the overpass must be 20 ft. above the roadway. What will the elevation of the bottom of the overpass be?

13. A roadway passing through a ravine has a vertical curve with an equation of:
$y = 0.0001x^2 - 0.08x + 10$.
a.) Plot the curve
b.) The county would like to put a pipe 8 ft. below the roadway at a location 300 ft. from where the vertical curve starts ($x = 300$). What will the elevation of the pipe be?

9.2 SOLVING QUADRATIC EQUATIONS FOR x

The preceding examples showed how to determine if a parabolic path will result in a particular outcome. Most problems involving quadratic equations require making the parabolic curve produce the results we want. In other words, instead of determining the value of y for a given value of x, we may need to find the value of x for a given value of y. For example, civil engineers might need to make the vertical curve of a new roadway miss an existing overpass. To do these types of problems, we need to be able to solve quadratic equations for x.

The first step in solving a quadratic equation for x is to *factor* the equation. To factor an equation is to express it in terms of the products of factors. Once this is done, a property known as the Zero Product Principle is used.

> The Zero Product Principal
>
> If $ab = 0$ then either $a = 0$ or $b = 0$
>
> or $a = b = 0$

Once the quadratic equation is factored, the solution is determined by finding the values of x which cause the factors to equal zero.

Example 9.5: In a volleyball game, Tim dives to save the ball 3 ft. from the net just before it hits the ground ($y = 0$). The top of the net is 8 feet above the ground. He hits the ball in the direction straight toward the opposite side on a trajectory of:

$y = -x^2 + 8x$ where x and y are in feet

a.) Does the ball make it over the net?

b.) If so, how far from the net will it land if unstopped?

Solution: a.) To determine if the ball makes it over the net, we need to find the height of the ball (y) at the net ($x = 3$).

at $x = 3$. $y = -x^2 + 8x$

$$= -(3^2) + 8(3) = 15 > 8 \text{ ft.}$$

The ball makes it over the net.

b.) To determine where the ball lands, we need to calculate the value of x when $y = 0$. The key to solving these types of problems is the Zero Product Principle. To be able to use the Zero Product Principal, we need to get the $-x^2 + 8x$ term into the ab form. In other words, we need to represent the term as the product of two factors. The $-x^2 + 8x$ term is factored out to determine how the factors can equal zero. We can see that an x will factor out so:

$$y = -x^2 + 8x$$

$$= x(-x + 8) = 0$$

In this equation, the x represents the a in the Zero Product Principal and the $(-x + 8)$ term represents the b. From the Zero Product Principal, we see that either $x = 0$, or $(-x + 8) = 0$. Therefore either $x = 0$ or $x = 8$. Written another way, $x \in \{0, 8\}$. This is said "x is an element of the set zero and eight."

Solving quadratic equations will always give two answers. In this example, the parabola intersects the x-axis ($y = 0$) at two places. It is up to you to determine which answer makes sense for the problem. The first solution was $x = 0$. By substituting 0 for x in the original equation, we can see that $y = 0$ at $x = 0$. This was when Tim first hit the ball. It hits on the other side of the net at $x = 8$ or $(8 - 3 = 5)$ 5 feet from the net on the other side.

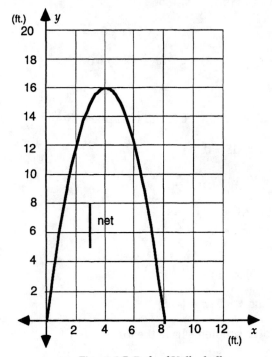

Figure 9.7 Path of Volleyball

The ball hits the ground 5 feet from the net on the other side.

A more common form of the quadratic equation is

$$x^2 + 5x + 6 = 0$$

This type of equation can also be solved by factoring but is somewhat more complicated. The solution will have the form:

$$(x + a)(x + b) = 0$$

If this is multiplied out, it looks like:

$$x^2 + ax + bx + ab = 0$$

or

$$x^2 + (a + b)x + ab$$

The solution to the problem is determined by finding two numbers, a and b, that multiply to 6 ($ab = 6$) and add to 5 ($a + b = 5$). One way to find these numbers is to list the factors of 6 and show their sum.

Factors	Σ
1 x 6	7
2 x 3	5 ←

The factors 2 and 3 work so the solution is:

$$(x + 2)(x + 3) = 0$$

Therefore, either $(x + 2) = 0$ or $(x + 3) = 0$

So $x = -2$ or $x = -3$ $x \in \{-3, -2\}$

Problems with negative factors are found the same way.

$$x^2 - 2x - 15 = 0$$

Factors	Σ
-1 x 15	14
1 x -15	-14
3 x -5	-2 ←
-3 x 5	2

$$(x + 3)(x - 5) = 0$$

Therefore $x = -3$ or $x = 5$ $x \in \{-3, 5\}$

Even more complicated problems have the form:

$$3x^2 - 18x + 24 = 0$$

$$(3x - 6)(x - 4) = 0$$

$$x = 2 \text{ or } x = 4$$

This type of problem could have been simplified by noticing that a 3 could be factored out of the equation first.

$$3(x^2 - 6x + 8) = 0 \qquad x = 2 \text{ or } x = 4 \qquad x \in \{2,4\}$$

As stated earlier, quadratic equations occur in numerous areas in the professional world. The following problem is an everyday example of how quadratic equations can occur.

Example 9.6: Sue and Bill would like to build a swimming pool in an 80 foot by 50 foot area in their backyard. From the contractor's estimates, they have determined they can afford 1,000 square foot pool. They would like to maintain a constant width around the pool between the pool and the perimeter of the area. How wide should the border be?

Solution: For problems like this, it is often helpful to draw a picture of the problem to help visualize the situation.

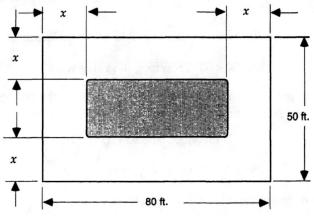

Figure 9.8 Pool Dimensions

x = the width of the border around the pool, in ft.

The key to solving this problem is finding the length of x that will create a pool with an area of 1,000 ft.^{2}

The area of the pool $\quad = \quad (50 - 2x)(80 - 2x) = 1,000$

$$2(25 - x)\, 2(40 - x) = 1000$$

$$\left(\frac{1}{4}\right) 4(25 - x)(40 - x) = \left(\frac{1}{4}\right) 1000$$

$$(25 - x)(40 - x) = 250$$

$$1000 - 65x + x^2 = 250$$

$$x^2 - 65x + 750 = 0$$

Here we need to find 2 numbers that add to -65 and multiply to 750. Again, we make a table.

Factors	Σ
-750 x -1	-751
-250 x -3	-253
-150 x -5	-155
-75 x -10	-85
-50 x -15	-65 ←

The equation is therefore factored as:

$$(x - 50)(x - 15) = 0$$

$$x = 50 \text{ or } x = 15 \qquad x \in \{15, 50\}$$

As usual, the solution to a quadratic equation gives us two answers. We need to go back to the original problem to determine which solution makes sense. We know that $2x$ has to be less than both 50 and 80.

$$2 \times 50 = 100 > 50 \qquad \text{Therefore 50 doesn't make sense.}$$

$$2 \times 15 = 30 < 50 \quad \text{Therefore, the value we want is 15 ft.}$$

It's a good idea to check the solution to make sure it works. If the border around the pool is 15 ft., its dimensions are:

$$80 - 2(15) = 50 \quad \text{and} \; 50 - 2(15) = 20$$

$$50 \times 20 = 1{,}000 \text{ ft.}^2 \quad \text{just as desired.}$$

The border between the pool and the fences should be 15 feet.

Exercise Set 9.2

1. Before a slam dunk, a basketball player jumps to the hoop with a vertical velocity of 40 ft./sec. By factoring Equation 9-4, determine his hang time.

2. A retractable cable supports a canopy at a right angle as shown in Figure A. In the open position, the cable extends 5 ft. When retracted, the cable extends 1 ft. Using the Pythagorean Theorem, determine the length of the canopy.

1 ft.

5 ft.

|◄— ? —►|

Figure A

3. A forest ranger would like to build an observation deck around her lookout tower as shown in Figure B. The dimensions of the tower are 10 by 20 feet. She has enough lumber to build 1,000 ft.^{2} of deck. If the deck has a constant width around the perimeter of the tower, what is the width?

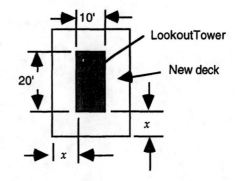

Figure B

Factor and determine the roots of the following equations:

4. $x^2 - 5x + 4 = 0$

5. $x^2 + 6x + 9 = 0$

6. $x^2 - 6x + 8 = 0$

7. $x^2 - 6x + 5 = 0$

8. $x^2 + 8x + 15 = 0$

9. $x^2 - 14x + 48 = 0$

10. $x^2 + 12x + 32 = 0$

11. $x^2 + 13x + 42 = 0$

12. $x^2 - 12x + 20 = 0$

13. $x^2 + 7x + 12 = 0$

14. $x^2 + 4x - 5 = 0$

15. $x^2 - x - 42 = 0$

16. $x^2 + 2x - 35 = 0$

17. $x^2 + 4x - 21 = 0$

18. $x^2 - 7x - 30 = 0$

19. $x^2 + 4x - 45 = 0$

20. $x^2 - 3x - 18 = 0$

21. $x^2 + 7x - 18 = 0$

22. $x^2 - 4x - 77 = 0$

23. $x^2 - x - 6 = 0$

24. $2x^2 + 7x + 6 = 0$

25. $3x^2 - 8x + 4 = 0$

26. $5x^2 - 13x - 6 = 0$

27. $4x^2 + 10x + 6 = 0$

9.3 THE QUADRATIC FORMULA

So far, we've solved all of our quadratic equations by factoring. When a quadratic equation:

$$ax^2 + bx + c = 0$$

has all of the a, b, and c factors or when the factors are decimals, solving by factoring can be extremely difficult. Fortunately a formula exists that can solve any quadratic equation. The formula is obtained by solving the standard quadratic equation for x.

$$ax^2 + bx + c = 0 \qquad \text{for } a \neq 0$$

$$\left(\frac{1}{a}\right)ax^2 + \left(\frac{1}{a}\right)bx + \left(\frac{1}{a}\right)c = 0\left(\frac{1}{a}\right)$$

$$x^2 + \frac{bx}{a} + \frac{c}{a} = 0$$

$$x^2 + \frac{bx}{a} = -\frac{c}{a}$$

$$x^2 + \frac{bx}{a} + \left(\frac{b}{2a}\right)^2 = \left(-\frac{c}{a}\right)\left(\frac{4a}{4a}\right) + \left(\frac{b}{2a}\right)^2$$

$$\left(x + \frac{b}{2a}\right)^2 = -\frac{4ac}{4a^2} + \left(\frac{b}{2a}\right)^2$$

$$\left(x + \frac{b}{2a}\right)^2 = \frac{b^2 - 4ac}{4a^2}$$

$$x + \frac{b}{2a} = \pm\sqrt{\frac{b^2 - 4ac}{4a^2}}$$

$$x + \frac{b}{2a} = \frac{\pm\sqrt{b^2 - 4ac}}{2a}$$

$$x = \frac{-b \pm \sqrt{b^2 - 4ac}}{2a} \qquad\qquad (9\text{-}10)$$

The term $\sqrt{b^2 - 4ac}$ in the quadratic formula is known as the *discriminant* of the equation. If you should happen to come across a problem where $4ac$ is greater than b^2, you will have to take the square root of a negative number. There are no real solutions for the square roots of negative numbers. To take the square root of a negative number, one must use what is known as *imaginary* numbers. Among other things, imaginary numbers make up a common part of electronics. To learn more about imaginary numbers, you'll need to take a more advanced course in algebra.

Example 9.7: While throwing the shot put, George releases it at a height of 1.9m and gives it an initial velocity vector of $9i + 9j$, with units in meters. How far does the shot put travel?

Solution: Since the range equation, Eq. 9-9, only applies to problems involving objects that begin and end their travel at the same height, it cannot be used for this problem. The problem can be solved instead by modifying Eq. 9-7 to account for the height at which the travel begins. If the travel begins at $y = 1.9$, then:

$$y = -\frac{g}{2v^2{}_{0x}}x^2 + \left(\frac{v_{0y}}{v_{0x}}\right)x + 1.9 \qquad \text{at } x = 0,\ y = 1.9$$

To find where the shot put hits the ground, we need to find the value of x for which $y = 0$. Recall that the problem gives the velocities in terms of metric units. We therefore need to use the metric value of g, 9.81 m/sec.^{2}. By substituting the coefficients of the initial velocity vector for v_{0x} and v_{0y}, the following quadratic equation is developed:

$$0 = -\frac{9.81}{2(9^2)}x^2 + \left(\frac{9}{9}\right)x + 1.9$$

$$= -0.06x^2 + x + 1.9$$

The a, b, and c terms are then put into the quadratic formula.

$$x = \frac{-b \pm \sqrt{b^2 - 4ac}}{2a}$$

$$= \frac{-(1) \pm \sqrt{(1)^2 - 4(-0.06)(1.9)}}{2(-0.06)}$$

$$= \frac{-1 \pm \sqrt{1 + 0.46}}{-0.12}$$

$$= \frac{-1 \pm \sqrt{1.46}}{-0.12}$$

$$= \frac{-1 \pm 1.2}{-0.12}$$

$$\frac{-1 + 1.2}{-0.12} = -1.7$$

$$\frac{-1 - 1.2}{-0.12} = 18.3$$

$$x \in \{-1.7, 18.3\}$$

As usual the quadratic formula produces two solutions. Since we're looking for a throwing distance, only the positive value makes sense. The significance of the negative value could be seen by drawing the parabolic path of the shot put with the $x = 0$ at the point where the shot put was released. If you were to continue the curve in the negative direction, the curve would cross the x-axis ($y = 0$) at -1.7.

The shot put travels 18.3 meters before hitting the ground.

Exercise Set 9.3

1. A punt is kicked from a height of 3 ft. with an initial velocity vector of $45i + 55j$, with units in ft./sec. If untouched, what is the horizontal distance the ball will travel in yards?

Figure A

2. Section 3.3 in the Weight Room Mechanics Chapter discussed how beams experiencing bending moments are in compression on one side and in tension on the other side. When steel bars are placed in concrete beams, the intent is for the steel to carry the tension loads while the concrete carries the compression loads. Since the concrete extends below the steel bars, only part of the concrete is in compression. The depth of the section which is in compression is determined by a quadratic equation. In one beam, the depth of the compression section, x is given by the equation $4x^2 + 18x - 252 = 0$, with units in inches. Given the equation, determine the depth x.

3. An underpass is being built below a railroad. The underpass will follow a vertical curve with an equation of $y = 0.0004x^2 - 0.1x + 20$, with units in ft. Ground borings in the area have determined that a hard flat layer of rock exists at an elevation of 12 ft.

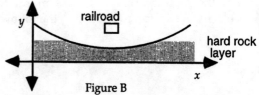

Figure B

The locations, x, where the road will meet the hard rock are therefore given by the equation:

$$12 = 0.0004x^2 - 0.1x + 18 \qquad \text{with units in ft.}$$

Determine the values of x where the hard rock will be reached. (i.e., at $y = 12$, $x = ?$)

4. A bridge spans a river which carries a lot of boat traffic. The bottom of the bridge has the equation:

$$y = -0.001x^2 + 0.1x + 34 \qquad \text{where } y \text{ is the}$$
height above the water level in ft.

a.) If the crest of the bridge is at $x = 50$, what is the height above the water of the tallest mast arm that could fit under the bridge?

b.) Most the boats that pass under the bridge have heights of 35 ft. What is the width of the space under the bridge in which these boats can travel?

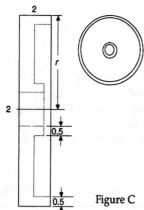

Figure C

5. Chapter 3 showed how to design some disc weights. You may have noticed that most weights in weight rooms are not simple flat plates with holes in them. A rim usually exists around the edge which makes the weight easier to handle. A rim is also usually provided around the hole to help keep the weights straight. A manufacturer wants to produce a 35-lb. weight with half-inch thick rims around the edge and hole. The disc will be made from a steel with a specific gravity of 0.3 lb./in.^{3}. The disc has a 2-inch diameter hole. The radius of the disc is determined by the equation:

$$0.3\pi\left[1\left(r^2 - 1^2\right) + 0.5\left(1.5^2 - 1^2\right) + 1\left(r^2 - (r - 0.5)^2\right)\right] = 35$$

which works out the following quadratic equation:

$$0.3\pi r^2 + 0.3\pi r - 35.6 = 0$$

Determine the required radius, r, of the disc.

Use the quadratic equation to solve the following equations for x:

6. $x^2 - 25 = 0$

7. $2x(x-1) = 3x + 42$

8. $x^2 + 3x - 28 = 0$

9. $x(x-3) = 4$

10. $-7x^2 - x + 60 = 0$

11. $8x^2 - 20x = 12$

12. $-x^2 + 5x - 6 = 0$

13. $x^2 - 4x + 3 = 0$

14. $3x^2 - x - 52 = 0$

15. $4x^2 + 10x - 6 = 0$

9.4 Parabolic Reflectors

In addition to punts and passes, parabolic curves sometimes appear in other places on the football field. TV and radio stations often like to include sound from the playing field in their broadcasts. Usually when people want to amplify sound, it's done with a microphone and a speaker. But a microphone on the sidelines would amplify sound coming from every direction and would therefore be useless. To focus in on sound coming from a particular location, a special property of the parabola is used. As their name implies, the parabolic mikes which are used on the sidelines are parabolic curves which have been rotated about an axis.

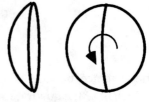

Fig. 9.9 Rotation of Parabola

Parabolic curves have the unique ability to focus parallel sound waves on a single point. They can also focus light rays and radio waves. The sound waves coming from a the field radiate out in all directions. By the time the sound waves reach the sidelines, they're traveling in nearly parallel waves. The parabolic mike focuses the sound waves coming from the field on a microphone where they're amplified electronically. TV satellite dishes are also parabolic and focus electromagnetic waves from satellites. Certain types of solar heaters are parabolic to focus sun rays on pipes which heat water or heat gases to run turbines for generating electricity.

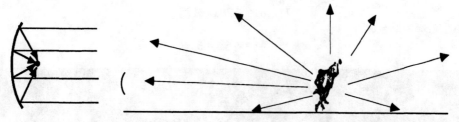

Figure 9.10 Concentration of Soundwaves

The focusing properties of the parabola also work in reverse. If a lightbulb is placed at the focus of a reflecting parabolic dish, all of the light that hits the dish will reflect out of the dish in parallel rays. This is how spot lights work.

To understand the focusing abilities of parabolic curves, we need to know the parts of a parabola. The curve itself is known as the *locus*. The point at which the waves or rays are focused is known as the *focus*. The center or end of the locus is known as the *vertex*. The locus is symmetrical about an axis known as the *axis of symmetry* which passes through the vertex. Below the locus and perpendicular to the axis of symmetry is a line known as the *directrix*. A fundamental property of the parabola is that every point on the locus remains the same distance from the focus as it is from the directrix.

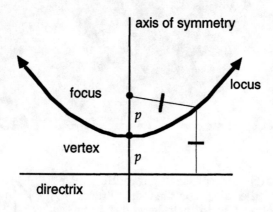

Figure 9.11 Parts of a Parabola

The distance between the focus and the vertex of the parabola depends on the a coefficient of the quadratic equation $ax^2 + bx + c = 0$. This distance is usually known as p. The relationship between p and the a coefficient is given by the equation:

$$p = \frac{1}{4a} \qquad\qquad (9\text{-}11)$$

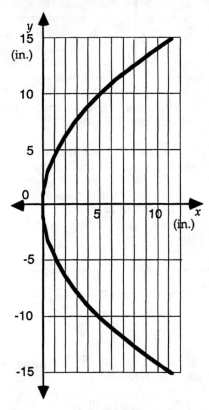

Fig. 9.12 Parabolic Mike Cross-section

Example 9.8: The microphone of a parabolic mike on the sidelines has been damaged by a stray linebacker. The microphone screws into a slot at the vertex. The parabola of the dish has an equation of:

$$x = \frac{1}{20}y^2 \qquad |y| \le 15 \quad \text{where the units are}$$

in inches.

How far should the new microphone be set from the vertex?

Solution: To fix the dish, we need to put the new microphone at the focus of the parabola. The required distance between the vertex and the focus is given by Equation 9.11:

$$p = \frac{1}{4a}$$

In the equation:

$$x = \frac{1}{20}y^2 \qquad a = \frac{1}{20}$$

Therefore

$$p = \frac{1}{4\left(\frac{1}{20}\right)}$$

$$= \frac{20}{4} = 5 \text{ in.}$$

The microphone must be 5 in. from the vertex.

Exercise Set 9.4

1. A parabola has an equation of $y = \frac{1}{16}x^2$.

a.) What are the coordinates of its focus?

b.) What is the equation of its directrix?

2. A parabolic reflector has its microphone positioned at the coordinates of (-6,0). The coordinates of its vertex are (0,0).

a.) What equation describes its locus?

b.) What equation describes its axis of symmetry?

3. As its name implies, a parabolic trough collector is a long trough-shaped solar collector with the cross-section of a parabola. With these types of collectors, a pipe is run down the trough at the focus of the parabola. Gases or fluids are run through the pipe to collect the heat from the sun.

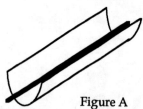

Figure A

If the parabola of a certain collector has the equation of $y = \frac{1}{48}x^2$, with units in inches, how far from the vertex should the pipe be placed?

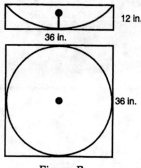

Figure B

4. A manufacturer of parabolic mikes is designing a new type of dish. Because of packing and shipping considerations, it would like the dish to fit in a box with inside dimensions of 12" x 36" x 36". How far from the vertex should the microphone be placed?

5. To collect a sufficient amount of radio waves a satellite dish manufacturer would like the outside rim to have a diameter of 48 in. It would like the focus of the parabola to be at the same height as the dish's rim as in Fig. C. What should the height of the dish be?

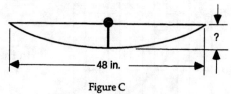

Figure C

6. A telephone company is designing a large parabolic reflector to receive and transmit phone calls from satellites. The equation of the parabola will be:

$$y = \frac{1}{40}x^2$$ with units in ft.

$$0 \le y \le 16$$

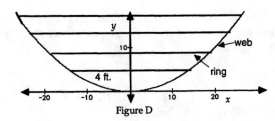

Figure D

The frame of the reflector will be made from a series of rings connected by webs as shown in Figure D. If the rings are to be spaced 4 ft. apart in the y-direction, what will their diameters need to be?

9.5 OTHER CONIC SECTIONS

The parabola is a member of a group of curves known as the *conic sections*. In addition to the parabola, the group includes:

circles
ellipses; and
hyperbolas

The group gets its name from the fact that the curves can be found by slicing cones at different angles. If two identical cones are placed point to point, each curve can be found by slicing one or both cones. If a slice is made perpendicular to the axis of a cone, a circle is created. If the slice is slightly tilted, an *ellipse* is created. If a slice is made parallel to one of the sides of the cone, a parabola is created. If a slice cut into both cones, as is shown in the figure, a *hyperbola* is created.

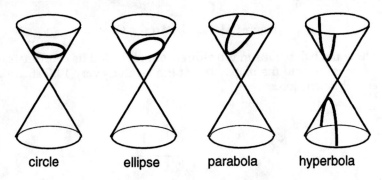

circle   ellipse   parabola   hyperbola

Figure 9.13 The Conic Sections

9.5.1 The Circle

The general equation for a circle centered at the origin with radius, r, is given by the equation:

$$x^2 + y^2 = r^2 \qquad\qquad (9\text{-}12)$$

The equation of the circle helps explain one of the most baffling phenomena in sports. When baseball players use the term "curve," they're probably referring to a breaking ball which is very hard to hit. A group of scientists once determined that the travel of curving pitches follows the path of a circular curve. If a pitch were given only side spin (like a slider or screwball), and neither hit any obstructions nor were affected by gravity, it would travel a full circle with a radius of about 2,000 feet. From the batter's perspective, the curve ball doesn't seem to break until the last few feet. It almost appears to turn a corner or drop off a table. By breaking down the motion into the xy components of the equation of a circle, we can better understand its travel. The equation of a circle with radius of 2,000 feet and centered at the origin is:

$$x^2 + y^2 = 2,000^2$$

(Note the similarities of the equation with the Pythagorean Theorem.)

To analyze the motion of the pitch, we'll say that the ball is released by a right handed pitcher at the point (2000,0). The ball travels about 60 feet in the positive y-direction.

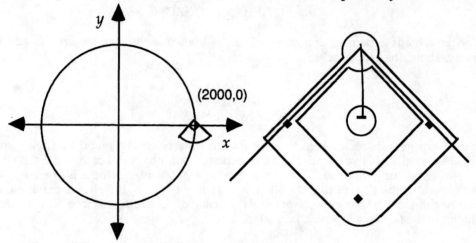

Figure 9.14 Circular Arc of a Curveball

To monitor how the ball breaks in the x-direction, we'll find its coordinates on the curve every 10 feet on its way to the plate. To get the x-value every 10 feet, we need to solve the equation of the circle for x.

$$x^2 + y^2 = 2,000^2$$

$$x^2 = 2,000^2 - y^2$$

$$x = \pm\sqrt{2,000^2 - y^2}$$

Since this example takes place entirely in the first quadrant (when x and y are positive), we can say:

$$x = \sqrt{2,000^2 - y^2}$$

Table 9.1 shows the movement of the pitch. The Δx (ft.) column represents the distance the pitch has traveled in the x-direction from the release of the pitch. The Δx (in.) column shows the same distance in inches. The $\Delta(\Delta x)$ column indicates how far the pitch has moved in the x direction in the past 10 feet.

y (ft.)	x (ft.)	Δx (ft.)	Δx (in.)	$\Delta(\Delta x)$ (in.)
0	2000.00	0.00	0.00	
10	1999.97	0.03	0.30	0.30
20	1999.90	0.10	1.20	0.90
30	1999.77	0.23	2.70	1.50
40	1999.60	0.40	4.80	2.10
50	1999.37	0.63	7.50	2.70
60	1999.10	0.90	10.80	3.30

Table 9.1 Displacements of a Curveball

As you can see in the table, the pitch breaks less than 3 inches in the first 30 feet. But by the time the ball is approaching the plate, it is breaking about 3 inches every 10 feet. It is this kind of travel which gives the ball the appearance of a drastic break. When the spin of the ball causes it to curve downward and is combined with the acceleration of gravity, it gives it the illusion of dropping off a table.

9.5.2 Horizontal Curves

A variation of Equation 9-12 exists for circles not centered at the origin. A circle with radius, r, and centered at a point with the coordinates (a,b) will have the equation:

$$(x - a)^2 + (y - b)^2 = r^2 \qquad\qquad (9\text{-}13)$$

The discussion of vertical roadway curves also made reference to *horizontal curves*. When roads or railroads curve to the left or right (as opposed to up or down) circular curves are used. When coordinate systems are used in the design of horizontal curves, the equation for circles is used.

Example 9.9: A team of engineers is in the process of widening a curved section of a road. The new inside edge of the road will have a radius of 100 ft. and will be centered at a point with the coordinates (10,25). A power line currently traverses the area on a line given by the equation $y = -1.2x + 169$. Because of the widening, the poles of the power line will need to be relocated to the edge of the new roadway. Determine the coordinates of the points at which the new poles will be located.

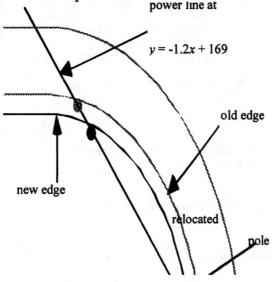

power line at

$y = -1.2x + 169$

old edge

new edge

relocated

pole

Solution: The poles will be relocated to the points where the line and the curve intersect. The problem states that the curve has a 100 ft. radius and that its center is (10,25). Using Eq. 9-13, its locus will be given by the equation:

$$(x-10)^2 + (y-25)^2 = 100^2$$

The poles will be located at the points where the curve and the power line intersect. The procedure used to find the intersection of a curve and a line is similar to that used to find the intersection of two lines. The solution can be found by substituting the value of y from the linear equation into the equation for the circular curve.

$$(x-10)^2 + (y-25)^2 = 100^2$$

$$(x-10)^2 + ((1.2x+169)-25)^2 = 100^2$$

$$(x-10)^2 + (1.2x+144)^2 = 10,000$$

$$x^2 - 20x + 100 + 1.44x^2 - 345.6x + 20,736 - 10,000 = 0$$

$$2.44x^2 - 365.6x + 10,836 = 0$$

The quadratic formula may then be used to determine the values of x:

$$x = \frac{365.6 \pm \sqrt{365.6^2 - 4(2.44)(10,836)}}{2(2.44)}$$

$$= \frac{365.6 \pm 167}{4.88}$$

$$x \in \{40.7, 109.1\}$$

The two numbers represent the x-values of the two points where the line intersects the circle. The values of y can then be determined by substituting the x-values into one of the other equations. The simplest equation to use is the linear equation.

$$y = -1.2(40.7) + 169 = 120.2$$

$$y = -1.2(109.1) + 169 = 38.1$$

The coordinates of the new locations for the poles are (40.7,120.2) and (109.1,38.1).

9.5.3 The Ellipse

Compared with the circle, the ellipse is a more complex curve. The standard equation for an ellipse centered at the origin is:

$$\frac{x^2}{a^2} + \frac{y^2}{b^2} = 1 \qquad\qquad (9\text{-}14)$$

Similar to the parabola, there are several different components which make up an ellipse. Figure 9.16 shows the different parts of an ellipse centered at the origin. Like a circle, an ellipse has a center. Two of the primary elements of an ellipse are its *major* and *minor* *axes*. The major axis extends the length of the curve in its long direction. The major axis can be either horizontal or vertical. The minor axis extends the length of the curve in its short direction. The lengths of the axes can be determined directly from an equation in the form of Ex. 9-14. The *a* and *b* terms in Eq. 9-14 represent one-half the length of the axis in the *x*- and *y*-directions respectively.

At the ends of the axes are the *vertices* of the ellipse. The coordinates of the four vertices of an ellipse are shown in Figure 9.16. Along the major axis are two *foci* (plural of focus).

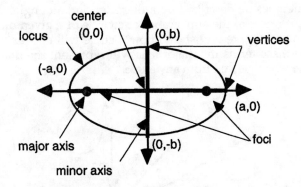

Figure 9.16 Parts of an Ellipse

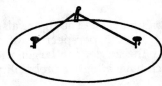

Fig. 9.17 Drawing an Ellipse

Similar to the parabola, the ellipse holds the property that any type of sound or light emanating from one of the foci will reflect off the surface of the ellipse and focus on the other focal point. Another property of the ellipse is that the locus represents the set of points of which the sum of the distances to the foci remains constant. A common way to describe this property is with the example of two pins stuck on a board with a loose piece of string tied between them. If a pencil were used to stretch the string as far as possible while drawing on the board, the resulting curve would be an ellipse. Since the length of the string cannot change, the sum of the distance from one focus to the pencil plus the distance from the pencil to the other focus remains constant.

The locations of the foci may be determined by the fact that distance between a vertex on the minor axis and a focal point equals *b* or half the length of the major axis. The relationship is shown in Fig. 9-18. In mathematical terms:

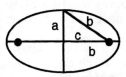

Fig. 9.18 Foci Locations $$c^2 = b^2 - a^2$$ (9-15)

A variation of Eq. 9-14 exists for ellipses not centered at the origin. An ellipse centered at the point (m,n) and having an axis length in the x-direction of a and an axis length of b in the y-direction will be given by the equation:

$$\frac{(x-m)^2}{a^2} + \frac{(y-n)^2}{b^2} = 1 \qquad (9\text{-}16)$$

An ellipse may sometimes be described by a diameter and an angle. When this type of description is used, it refers to the appearance of a circle of a given diameter when view from a particular angle. Figure 9.19 shows how a circle will appear when viewed from different angles. In these terms the major axis will represent the diameter of the circle and remain the same when viewed from any angle.

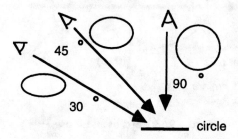

Figure 9.19 Ellipses by Degrees

Figure 9.19 has a special significance for basketball players. The figure also indicates how the rim of a basket appears from the perspective of a basketball approaching the basket. The greater the angle, the larger the target. This explains why shots with greater arch have a higher probability of being successful.

9.5.4 The Hyperbola

Similar to the ellipse, the hyperbola represents the set of points which maintain a constant difference in the distances between focal points. The general equation for a hyperbola centered at the origin is:

$$\frac{x^2}{a^2} - \frac{y^2}{b^2} = 1 \qquad (9\text{-}17)$$

The hyperbola is made up of two curves that are similar in shape to a parabola. Like the parabola, the two curves are centered on an axis. A hyperbola can also have a vertical orientation. If the axis is horizontal, the vertices of the curves will be a distance a from the center. If the axis is horizontal, the vertices will be a distance b from the center.

The ends of the curves approach two lines that are known as *asymptotes*. While the curves get very close to the asymptotes, they never actually reach them. The curves are said to approach the lines *asymptotically*. The equations for the asymptotes are:

$$y = \frac{b}{a}x \qquad \text{and} \qquad y = -\frac{b}{a}x \qquad (9\text{-}18)$$

Figure 9.20 shows a hyperbola with the equation $\frac{x^2}{3^2} - \frac{y^2}{4^2} = 1$.

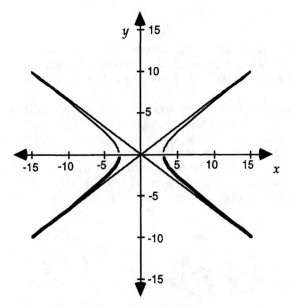

Figure 9.20 Hyperbola

A different type of hyperbola has the form:

$$y = \frac{k}{x} \qquad \text{where } k = \text{ a constant} \qquad (9\text{-}19)$$

In this type of hyperbola, the x- and y-axes are the asymptotes. For example the equation $y = \frac{5}{x}$ is shown in Figure 9.21:

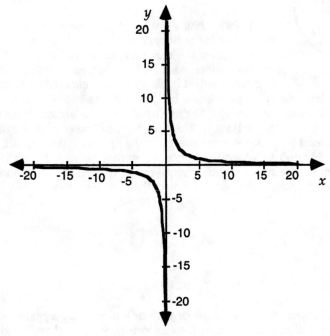

Figure 9.21 Hyperbola

9.5.5 Orbits

This chapter began by discussing how gravity causes an object traveling through the air to travel in a parabolic curve. Under certain conditions in space, gravity can also cause objects to travel in the paths of the curves of the other conic sections.

The physics of an object traveling in an orbit are somewhat different from one traveling near the earth's surface. In space, of course, there is no air friction to affect the path of an object as in the earth's atmosphere. Objects in space, therefore, more closely follow the paths of the conic sections. Unless the object is physically hit, a path caused by the gravitational force of one planet or sun can only be changed by the gravitational force of another. Near the earth's surface, the source of the gravitational force is effectively a plane. In orbit, the source of gravity is actually a point at the center of the earth.

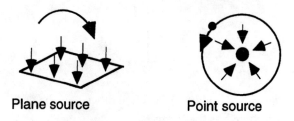

Plane source   **Point source**

Figure 9.22 Gravity Sources

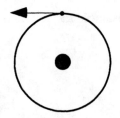

Fig. 9.23 Circular Orbit

If an object has a certain amount of kinetic energy, it will travel in a circular orbit. Since it remains the same height above the earth's surface, it neither gains potential energy nor loses its kinetic energy.

If an object has a little more kinetic energy, it will begin to pull out of its circular orbit and into an elliptical orbit. The ellipse will have the center of the earth as one of its foci. Just like a ball near the earth's surface, as the object begins to pull away, the earth's gravity begins to pull it back. As it travels farther from the earth, it begins to gain potential energy. Again like a ball, the object's movement away from the earth is eventually overcome by gravity. At the ellipse's vertex farthest from the earth (called the *apogee*), it begins to travel back in the direction of the earth. At this point, its kinetic energy is the least and its potential energy is the greatest. As a result, the object begins to accelerate at this point. When it reaches the vertex nearest the earth (the *perigee*), it will have its greatest amount of kinetic energy but will have the least potential energy. With its kinetic energy increased again, the object then begins a new cycle.

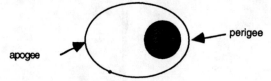

apogee   perigee

Fig. 9.24 Elliptical Orbit

Sometimes an object such as a comet or meteor will pass by a planet with so much kinetic energy that it will not be brought into an orbit. Its travel will still be affected by the gravity of the planet. If this occurs, there is one speed at which the object will travel in the path of a parabola. Unlike the parabolic travel of an object near the earth's surface, the maximum speed of a parabolic orbit occurs at the vertex of the parabola. At any speed greater than this, it will travel in the path of a hyperbola.

Exercise Set 9.5

1. A ping pong player puts enough "english" on the ball to give it a curve with a radius of 100 ft. from the position shown in Figure A.

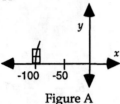

Figure A

How far will the ball curve in 8 ft.? (i.e., $\Delta x = ?$ for $\Delta y = 8$)

2. A basketball player shoots from the 3-point line giving the ball a horizontal trajectory described by the equation:

$$y = -\frac{1}{2}x + 20.$$

The equation of the 3-point line is $x^2 + (y - 20)^2 = 324$. What are the coordinates of the location from which the shot was made?

3. A team of engineers is designing the horizontal curve of a new roadway. A drainage pipe passes under the area where the horizontal curve is planned.

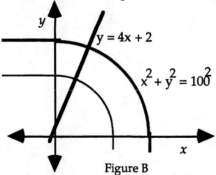

Figure B

The engineers wish to put a drain at the outside curb of the roadway above the drain pipe. Relative to the center of the curve, the pipe is located along the line $y = 4x + 2$. The outside curb of the horizontal curve has the equation of:

$x^2 + y^2 = 10,000$ for $0 \le x \le 100$, and $0 \le y \le 100$. Determine the coordinates of the drain.

4. An astronomer is plotting what appears to be the elliptical orbit of a planet in a distant galaxy. Since there is a cluster of stars located where the focus of the ellipse should be, it is not apparent around which star the planet is orbiting.

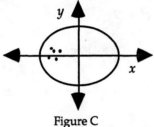

Figure C

After doing some calculations, she develops a coordinate system for the ellipse's vertices. Three of the vertices have the coordinates of (0,30.2), (46.3,0), and (0,-30.2). Given this coordinate system, determine the coordinates of the star around which the planet is orbiting.

5. The cross-section of a new blimp will consist of the halves of two different ellipses connected together as shown in Figure D. The ellipses will have minor axes of the same lengths but the lengths of their major axes will differ.

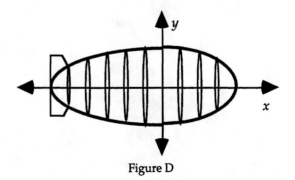

Figure D

The equations of the ellipses are:

$$\frac{x^2}{3600} + \frac{y^2}{400} = 1 \qquad \frac{x^2}{1600} + \frac{y^2}{400} = 1$$

$$-60 \le x \le 0 \qquad\qquad 0 \le x \le 40$$

The frame of the blimp will consist of large rings spaced 10 ft. apart in the horizontal direction. Determine the radii of the rings required to construct the blimp.

6. Using Eq. 9-9, plot the combinations of initial horizontal and vertical velocities which will result in an object traveling 30m.

Plot the following curves:

7. $x^2 + y^2 = 25$

8. $x^2 + y^2 = 17$

9. $(x-3)^2 + (y+4)^2 = 16$

10. $(x+7)^2 + y^2 = 9$

11. $\dfrac{x^2}{9} + \dfrac{y^2}{49} = 1$

12. $\dfrac{x^2}{5} + \dfrac{y^2}{1} = 1$

13.

$$\frac{(x-4)^2}{16} + \frac{(y+6)^2}{36} = 1$$

14.

$$\frac{(x+2)^2}{20} + \frac{(y-5)^2}{8} = 1$$

15. $\dfrac{x^2}{4} - \dfrac{y^2}{9} = 1$

16. $\dfrac{y^2}{16} - \dfrac{x^2}{25} = 1$

17. $xy = 24$

18. $xy = -36$

REFERENCES

E. Schrier, W. Allman, *Newton at the Bat: The Science in Sports*, Charles Scribner's Sons, New York, 1987. The research and documentation of the circular travel of breaking pitches.

R. Brinker, P. Wolf, *Elementary Surveying*, Harper & Row, New York, 1977. Discussions on vertical and horizontal railroad and roadway curves.

Chapter 10

POLYNOMIALS OPERATIONS

So far in this book, we've looked at polynomial equations which represent formulas for subjects including material mechanics, statistics, linear motion, and parabolic travel. The equations shown in this book represent only a small fraction of those that can be found in the professional world. The equations in business, agriculture, biology, and engineering can take almost any form imaginable. Some highly complex examples of the polynomial equations found in the professional world include:

$$P = \frac{(1+i)^n - 1}{i^2 (1+i)^n} - \frac{n}{i(1+i)^n}$$ a financial formula

$$P_x = \frac{d^2_{A-x} P_A + d^2_{B-x} P_B + d^2_{C-x} P_C + d^2_{D-x} P_D}{d^2_{A-x} + d^2_{B-x} + d^2_{C-x} + d^2_{D-x}}$$ a formula for estimating rainfall

$$Q = \sqrt{\frac{2g(h_1 - h_4)}{1 + \frac{29C^2_d L}{(r)^{\frac{4}{3}}}}}$$ a formula for designing waterways

To be able to work with these types of equations, algebra students need to be able to handle polynomials of virtually any form. Just as arithmetic students need to be able to add, subtract, multiply, and divide numbers, algebra students must be able to add, subtract, multiply, and divide polynomials. Many of the rules for working with polynomials have already been shown. For example, the Associative and Commutative properties of addition and multiplication also apply to polynomials. This chapter shows some additional techniques for working with polynomials.

10.1 ADDITION AND SUBTRACTION OF POLYNOMIALS

The addition and subtraction of polynomials involves collecting *like terms* of a polynomial and adding or subtracting their coefficients. Of the polynomials:

$$4a^2 - 5x^3 \sqrt[4]{y} \quad \text{and} \quad 2x^3 \sqrt[4]{y} + 2a^2 b,$$

$$\text{the terms} \quad -5x^3 \sqrt[4]{y} \quad \text{and} \quad 2x^3 \sqrt[4]{y}$$

are considered to be like terms. This is because the terms have the same variables and the variables have the same exponents. The sum of these two polynomials is calculated as follows:

$$4a^2 - 5x^3\sqrt[4]{y}$$
$$+ \qquad 2x^3\sqrt[4]{y} + 2a^2b$$
$$\overline{4a^2 - 3x^3\sqrt[4]{y} + 2a^2b}$$

If the second polynomial were subtracted from the first, the difference would be calculated as follows:

$$4a^2 - 5x^3\sqrt[4]{y}$$
$$- \qquad 2x^3\sqrt[4]{y} + 2a^2b$$
$$\overline{4a^2 - 7x^3\sqrt[4]{y} - 2a^2b}$$

When possible, it is standard practice to arrange polynomials with a descending order of exponents. For example, if a problem involving polynomials produced the following answer:

$$2x^2 - 3x^5 + 7 + 4x^3 - 6x^4$$

it would usually be rearranged to the following order:

$$-3x^5 - 6x^4 + 4x^3 + 2x^2 + 7$$

Note that the exponents descend from left to right. Arranging polynomials by exponents assists if factoring is necessary. If the terms of a polynomial include two or more variables and a significant pattern would not result from ordering the terms, the practice is usually not done.

Exercise Set 10.1

Add the following polynomials:

1.
$$2x^3 + 4x^2 - 7x + 17$$
$$\underline{-x^4 + x^2 - 23}$$

3.
$$2x^2 - 7x - 3xy + 4y - y^2$$
$$\underline{-5y^2 + xy - 2x - 6x^2}$$

5.
$$x^{\frac{3}{2}} + 5x^{\frac{1}{2}} - x^{-\frac{1}{2}}$$
$$\underline{-x^{\frac{3}{2}} - 2x^{\frac{1}{2}} + 3x^{-\frac{1}{2}} + 8x^{-\frac{3}{2}}}$$

2.
$$3xy + 7y^2 - 8x^2 + 27$$
$$\underline{5x - 9y + y^2 - 18}$$

4.
$$3\left(x^2 - xy + y^2\right) + x - y$$
$$\underline{5xy - xy^2 - 8y^2 + 4}$$

6.
$$3x^3 - 5x - 12$$
$$\underline{6x^2 - x + 2x^{\frac{1}{2}}}$$

Subtract the following polynomials:

7.
$$2x\sqrt[3]{y} + 7$$
$$x^3 - 3x\sqrt[3]{y} + x^2$$

9.
$$x^2 - 25$$
$$4\left(y^2 + x^2\right) - 3xy + 6$$

11.
$$14x^5 - 6x^2$$
$$-8x^5 + 7x^3 + 3x^2 - 6x + 4$$

8.
$$7y + 3x\sqrt[5]{y} - 4$$
$$y^2 - 2xy + 2\sqrt[3]{y} + 2x\sqrt[5]{y}$$

10.
$$5x - 2xy + 7y + 13$$
$$6\left(x^2 + xy + y^2\right) - 4$$

12.
$$13x^2y^2 + 2x^2 - 3y^2 + 6$$
$$6x^2 - x - 9x^2y^2 + 7y$$

10.2 MULTIPLICATION OF POLYNOMIALS

The procedure used to multiply polynomials is based on the Distributive Property of Multiplication over Addition. If the following two binomials needed to be multiplied:

$$(a + b)(c + d)$$

one could assign a variable f to represent $c + d$. In other words $f = c + d$. The equation could then be written as:

$$(a + b)f$$

Using the Distributive Property, this term equals $af + bf$.

If $f = c + d$ the term would equal $a(c + d) + b(c + d)$.

The Distributive Property is then used again to generate the answer

$$ac + ad + bc + bd$$

Mathematicians use an acronym, FOIL, to remember how to multiply binomials. The technique is to first multiply the two First terms in each factor, then the two Outside terms, the two Inside terms, and finally the two Last terms. Visually, the technique looks as follows:

$$\text{(a + b)(c + d)} = ac + ad + bc + bd$$

F   O   I   L

Note that this is the reverse of the process used to factor quadratic equations. A similar technique is used when a factor contains more than two terms. In this case, each term of a factor must be multiplied by every other term in the other factor. For example:

$$(a + b)(c + d + e) = ac + ad + ae + bc + bd + be$$

When polynomials are multiplied, the product may yield terms that may be combined. For example, in the problem:

$$\left(2x - 3\sqrt{x}\right)\left(4\sqrt{x} + 5\right) = 8x\sqrt{x} + 10x - 12x - 15\sqrt{x}$$

the terms $10x$ and $-12x$ can be added. When the like terms are combined, the answer is:

$$8x\sqrt{x} - 2x - 15\sqrt{x}$$

Exercise Set 10.2

Expand the following polynomials:

1. $(x - 4)(x + 7)$

3. $\left(x^2 - 7\right)(x + y)$

5. $\left(x^{\frac{3}{2}} - x\right)\left(x^{\frac{1}{2}} + 4\right)$

2. $x(x + 2)(x + 3)$

4. $(x + 2)\left(x^2 - 3x + 5\right)$

6. $\left(2x^2 + 6x - 3\right)\left(-x^2 + x - 4\right)$

10.3 DIVISION OF POLYNOMIALS

When polynomials appear in the numerators and denominators of fractions, the procedures used to reduce the fractions are similar to those used to reduce fractions with numbers in the numerator and denominator. The key to reducing fractions is to factor the numerator and denominator. For example, the fraction below may be factored and reduced as follows to produce the answer.

$$\frac{315}{231} = \frac{3 \cdot 3 \cdot 5 \cdot 7}{3 \cdot 7 \cdot 11} = \frac{15}{11}$$

Similarly, the polynomial fraction below could be factored and reduced as follows:

$$\frac{x^2 - 4x}{x^2 - x - 12} = \frac{x(x - 4)}{(x + 3)(x - 4)} = \frac{x}{x + 3} \qquad x \notin \{-3, 4\}$$

Note that it is necessary to exclude values which would result in division by zero and create unreal numbers. Values must be excluded not only for numbers in the answer that cause division by zero but also for values which would cause division by zero in any step of the problem.

Exercise Set 10.3

Simplify the following polynomial fractions:

1. $\dfrac{x^2 - 4x - 21}{x^2 + 9x + 18}$

3. $\dfrac{x^3 - x}{x^2 - x}$

5. $\dfrac{x^3 - 5x^2 + 6x}{x^3 - 7x^2 - 4x + 120}$

2. $\dfrac{x^2 + 5x + 4}{x^2 - x - 2}$

4. $\dfrac{x^2 + 3x - 40}{x^2 - 4x - 45}$

6. $\dfrac{x^2 + 10x + 16}{x^2 - 3x - 10}$

10.4 SPECIAL PRODUCTS

Since factoring is so important in working with polynomials, students should be able to recognize of some special patterns of polynomials and know how they can be factored. The following equations show some special polynomial products:

$$a^2 - b^2 = (a + b)(a - b) \qquad (10 \text{-} 1)$$

$$a^3 \pm b^3 = (a \pm b)\left(a^2 \mp ab + b^2\right) \qquad (10 \text{-} 3)$$

$$(a \pm b)^2 = a^2 \pm 2ab + b^2 \qquad (10 \text{-} 2)$$

$$(a \pm b)^3 = \left(a^3 \pm 3a^2b + 3ab^2 \pm b^3\right) \qquad (10 \text{-} 4)$$

$$\left(a^n + b^n\right) = (a + b)\left(a^{n-1} - a^{n-2}b + \cdots + b^{n-1}\right) \qquad \text{for n odd only} \qquad (10 \text{-} 5)$$

$$\left(a^n - b^n\right) = (a - b)\left(a^{n-1} + a^{n-2}b + \cdots + b^{n-1}\right) \qquad \text{for n odd or even} \qquad (10 \text{-} 6)$$

Equation 10-1 is usually referred to as the *difference of two squares*. Equation 10-3 shows how the difference and sum of two cubes can be factored.

<u>Example 10.1:</u>   Simplify

$$\frac{n^3 + m^3}{m^2 - n^2}$$

<u>Solution:</u> If the terms in the numerator of the fraction are reordered, the pattern becomes more visible. The fraction has the sum of two cubes in the numerator and the difference of two squares in the denominator. If we recognize these terms and know how they can be factored, we'll know that an $m + n$ term can be factored out of both the numerator and denominator.

$$\frac{n^3 + m^3}{m^2 - n^2} = \frac{m^3 + n^3}{m^2 - n^2} = \frac{(m + n)\left(m^2 - mn + n^2\right)}{(m + n)(m - n)} = \frac{\left(m^2 - mn + n^2\right)}{(m - n)} \qquad m \neq n, \quad m \neq -n$$

Note that the $m^2 - mn + n^2$ differs from the $a^2 - 2ab + b^2$ term in Eq. 10-2. The answer therefore cannot be factored any further.

Exercise Set 10.4

Use the identities to simplify the following polynomial fractions:

1. $\dfrac{x^2 - 4}{x^3 + 8}$

3. $\dfrac{u^4 - v^4}{u^3 - v^3}$

5. $\dfrac{x^3 + 9x^2 + 27x + 27}{x^2 + 6x + 9}$

2. $\dfrac{x^4 + 27x}{x^3 - 9x}$

4. $\dfrac{x^3 - 8}{x^4 - 16}$

6. $\dfrac{x^2 - 10x + 25}{x^2 - 25}$

10.5 POLYNOMIAL FRACTIONS

Operations with polynomial fractions are done in the same way as numbers. Again, when dealing with polynomial fractions, one must always remember to avoid operations, such as division by zero, which would create unreal numbers.

10.5.1 Multiplication

Polynomial fractions are multiplied in the same way as number fractions. The numerators of the fraction factors are multiplied together as are the denominators.

$$\left(\frac{a}{b}\right)\left(\frac{c}{d}\right) = \frac{ac}{bd} \qquad b \neq 0, \quad d \neq 0$$

<u>Example 10.2:</u>   Calculate: $\left(\dfrac{1}{2}\right)\left(\dfrac{3}{4}\right)$

<u>Solution:</u> $\left(\dfrac{1}{2}\right)\left(\dfrac{3}{4}\right) = \dfrac{1 \times 3}{2 \times 4} = \dfrac{3}{8}$

<u>Example 10.3:</u>   Calculate: $\left(\dfrac{3x - 2y^2}{4x + y}\right)\left(\dfrac{y + 5}{z}\right)$

<u>Solution:</u> The numerators of the two fractions are multiplied using the FOIL process. The denominators are multiplied using the Distributive Property.

$$\left(\frac{3x - 2y^2}{4x + y}\right)\left(\frac{y + 5}{z}\right) = \frac{3xy + 15x - 2y^3 - 10y^2}{4xz + yz} \qquad z \neq 0, y \neq -4x$$

10.5.2 Division

Like numbers, polynomial fractions are divided by multiplying by reciprocals.

$$\frac{\dfrac{a}{b}}{\dfrac{c}{d}} = \left(\frac{a}{b}\right)\left(\frac{c}{d}\right)^{-1} = \left(\frac{a}{b}\right)\left(\frac{d}{c}\right) = \frac{ad}{bc} \qquad b \neq 0,\ c \neq 0,\ d \neq 0$$

<u>Example 10.4:</u> Simplify:

$$\frac{\dfrac{3}{5}}{\dfrac{9}{10}}$$

<u>Solution:</u>
$$\frac{\dfrac{3}{5}}{\dfrac{9}{10}} = \left(\frac{3}{5}\right)\left(\frac{9}{10}\right)^{-1} = \left(\frac{3}{5}\right)\left(\frac{10}{9}\right) = \frac{(3)(2)(5)}{(5)(3)(3)} = \frac{2}{3}$$

<u>Example 10.5:</u> Simplify:

$$\frac{\dfrac{6x-3}{y-1}}{\dfrac{4x-2}{y^2-1}}$$

<u>Solution:</u>

$$\frac{\dfrac{6x-3}{y-1}}{\dfrac{4x-2}{y^2-1}} = \left(\frac{6x-3}{y-1}\right)\left(\frac{4x-2}{y^2-1}\right)^{-1} = \left(\frac{6x-3}{y-1}\right)\left(\frac{y^2-1}{4x-2}\right)$$

$$= \left(\frac{3(2x-1)}{y-1}\right)\left(\frac{(y+1)(y-1)}{2(2x-1)}\right) = \frac{3y+3}{2} \qquad x \neq \frac{1}{2},\ y \notin \{-1,1\}$$

<u>Example 10.6:</u> Simplify:

$$\frac{x^3 + 9x^2 + 23x + 15}{x^2 - 6x - 27}$$

Although the denominator in this problem can be factored by inspection $(x + 3)(x - 9)$, the numerator is too large to be factored by inspection. Since all of the terms in the numerator are positive and the last term is 15 (whose only factors are 1, 3, 5, and 15), a good guess would be try to divide the whole term by $x + 5$. This can be done by long division in the same way that constants are.

$$\begin{array}{r} x^2 + 4x + 3 \\ x+5\overline{\smash{)}x^3 + 9x^2 + 23x + 15} \\ \underline{x^3 + 5x^2} \\ 4x^2 + 23x \\ \underline{4x^2 + 20x} \\ 3x + 15 \\ \underline{3x + 15} \\ 0 \end{array}$$

It turns out that $x + 5$ is a factor of the numerator.

$$x^3 + 9x^2 + 23x + 15 = (x+5)(x^2 + 4x + 3)$$

Now the factor on the right can be further factored.

$$x^2 + 4x + 3 = (x+1)(x+3)$$

The original fraction can then be written as:

$$\frac{(x+5)(x+1)(x+3)}{(x+3)(x-9)}$$

The $x + 3$ terms cancel out.

$$\frac{x^2 + 6x + 5}{x - 9} \qquad x \notin \{-3, 9\}$$

10.5.3 Addition and Subtraction

Addition and subtraction of polynomial fractions, like number fractions, require having common denominators. When the fractions to be added or subtracted do not have common denominators, they must be converted to create them.

Example 10.7:   Calculate

$$\frac{4}{5} + \frac{5}{21}$$

Solution: The first step to getting common denominators is to factor the denominators of both fractions to determine the *Least Common Denominator* or LCD.

$$\frac{4}{5} + \frac{5}{21} = \frac{4}{3 \cdot 5} + \frac{5}{3 \cdot 7}$$

From the equation, we can see that if they are to have the same denominator, the first term needs a factor of 7 and the second term needs a factor of 5. We get this by multiplying the fractions by equivalencies of 1.

$$\frac{4}{3\cdot5}\left(\frac{7}{7}\right)+\frac{5}{3\cdot7}\left(\frac{5}{5}\right)=\frac{28}{105}+\frac{25}{105}=\frac{28+25}{105}=\frac{53}{105}$$

The numerator, 53, has no factors other than itself and 1. Since it cannot be factored any further, it is known as a *prime* number.

The LCD of polynomial fractions must also be found before they can be added or subtracted.

<u>Example 10.8:</u>   Calculate:

$$\frac{3}{x^2-2x-8}-\frac{2}{x^2-4x}$$

<u>Solution:</u> The LCD is again found by factoring the denominators.

$$\frac{3}{x^2-2x-8}-\frac{2}{x^2-4x}=\frac{3}{(x+2)(x-4)}-\frac{2}{x(x-4)}$$

By inspection, we can see that to add the two fractions, the first term needs an x factor in the denominator and the second fraction needs an $(x+2)$ factor in the denominator. The terms are then multiplied by the appropriate equivalencies of one to create common denominators.

$$\frac{3}{(x+2)(x-4)}\left(\frac{x}{x}\right)-\frac{2}{x(x-4)}\left(\frac{x+2}{x+2}\right)=\frac{3x-(2x+4)}{x(x-4)(x+2)}=\frac{x-4}{x(x-4)(x+2)}$$

$$=\frac{1}{x(x+2)}=\frac{1}{x^2-2x}\qquad x\notin\{-2,0,4\}$$

Exercise Set 10.5

Simplify the following polynomials:

1. $\left(\dfrac{x^2 - x}{x+4}\right)\left(\dfrac{x-7}{x-1}\right)$

4. $\dfrac{\dfrac{x^2+4}{\sqrt{x-5}}}{\dfrac{x-5}{x-7}}$

7. $\dfrac{x}{x-7} - \dfrac{7}{x+1}$

2. $\left(\dfrac{x^3+8}{x^4}\right)\left(\dfrac{x+3}{x+2}\right)$

5. $\dfrac{x+7}{x^2-9x+20} + \dfrac{x+7}{x^2-7x+12}$

8. $\dfrac{x+3}{x^2-4x} + \dfrac{x+7}{x^2+2x}$

3. $\dfrac{\dfrac{x}{x-1}}{\dfrac{y+5}{y-1}}$

6. $\dfrac{x^2-4x}{x-6} - \dfrac{2x-24}{6-x}$

Chapter 11

LOGARITHMS

This chapter demonstrates how to solve equations entirely different from anything shown so far in this book. While the properties of exponents have been discussed, we have not yet placed the variables in the exponent. Sometimes problems may arise which take the form:

Solve for x:

$$100 = 10^x$$

While the answer of 2 might be found by inspection, a more difficult problem might have the form:

Solve for x:

$$5 = 2.73^x$$

To solve problems like this, we need to gain an understanding of the concept of *logarithms*. Like most math, logarithms have a number of different uses. The fundamental thing to remember about logarithms is that *logarithms are exponents*. There are generally two types of logarithms, common logarithms and natural logarithms.

11.1 COMMON LOGARITHMS

One of the most common uses for logarithms is the multiplication of large numbers. Computers and calculators use logarithms to multiply. The basis of the multiplication procedure is the exponent property:

$$a^m \cdot a^n = a^{m+n}$$

In this property, a is the *base* of the two terms a^m and a^n, and m and n are the exponents of the two terms.

Near the end of the 16th century, a Scottish mathematician named John Napier noticed that it is much easier to add large numbers than it is to multiply them. For example, if someone needed to multiply the two numbers 315 and 684, it would be much easier if you could convert the two numbers to numbers with exponents and a common base, for example 10, and then just add the exponents. In other words,

if $315 = 10^m$ and $684 = 10^n$,

then $315 \times 684 = 10^m \times 10^n = 10^{m+n}$

The problem is finding the exponents of 10 which cause it to equal these values. If the exponents for all numbers could be found and tabulated, the multiplication of large numbers could be simplified to a process of addition. All one would need to do to multiply large numbers would be to look up the values of these exponents or logarithms, add them and then find the number whose exponent corresponds to this sum.

The process is better illustrated with smaller numbers. Table 11.1 shows the numbers from 1 to 10 in the top row and their logarithms, or *logs*, to the base 10 in the bottom row. In other words, we can see from the Table 11.1 that:

$$2 = 10^{0.30}$$

	1	2	3	4	5	6	7	8	9	10
logarithm	0.00	0.30	0.48	0.60	0.70	0.78	0.85	0.90	0.95	1.00

Table 11.1 Common Logarithms to Two Places

If we want to multiply the numbers 2 and 4, we look up their corresponding logs 0.30 and 0.60 and then add them. The sum of the two logs is 0.90. We then look up the *antilogarithm* or *antilog* of 0.90 and see that it is 8. Obviously $2 \times 4 = 8$. In exponential terms:

$$2 \times 4 = 10^{0.30} \times 10^{0.60} = 10^{0.30+0.60} = 10^{0.90} = 8$$

An equivalent process works for division. To divide 10 by 5, the log of 10, 1.00, has the log of 5, 0.70, subtracted from it. The difference is 0.30. The antilog of 0.30 is 2. In exponential terms:

$$\frac{10}{5} = \frac{10^{1.00}}{10^{0.70}} = 10^{1.00} \times 10^{-0.70} = 10^{1.00-0.70} = 10^{0.30} = 2$$

The base of a log is denoted by a subscript. The log of x to the base 10 is denoted as $\log_{10}x$. The properties for multiplication and division with logs are expressed by the following equations:

$$\log_a M \cdot N = \log_a M + \log_a N$$

$$\log_a \frac{M}{N} = \log_a M - \log_a N$$

The logs in Table 11.1 are accurate only to two decimal places. Logarithms accurate to three decimal places are shown in Appendix A. When computers and calculators multiply and divide, they use logs accurate to many more places. For example, a spreadsheet program will often use logs accurate to 15 digits.

While using logarithms may seem to be a difficult way to multiply and divide, it makes more sense for larger numbers. If the number 10 is used as the base, finding the exponents of large numbers is greatly simplified. Notice that:

$$315 = 3.15 \times 100$$

$$= 3.15 \times 10^2$$

Similarly,

$$0.315 = \frac{3.15}{10} = 3.15 \times 10^{-1}$$

$$\text{If } 10^n = 3.15 \quad \text{then} \quad 3.15 \times 10^2 = 10^n \times 10^2 = 10^{n+2}$$

Given that it is so easy to factor out the multiples of 10, we only need to find the logarithms from 1.0 to 10 to cover all numbers. Larger or smaller numbers can be accounted for by adding or subtracting the appropriate factors of 10.

The logarithms for positive numbers less than 1 or greater than 10 are often broken up into two parts to make them easier to multiply and divide. The first part of the logarithm is the decimal part or *mantissa*. The mantissa describes the number's magnitude between 1 and 10. The second part of the log is an integer known as the *characteristic*. The characteristic describes the number's magnitude as a factor of ten.

To determine the mantissa and characteristic of a number, it is first converted into exponential notation. For example to find the log of 248, the exponential form is 2.48 x 10^2. The logarithm of the mantissa is read from the tables in Appendix A by first finding the page with 2.4 on the row on the left side of the table. You then read across to the 0.08 column to find the logarithm value of 0.394. To three decimal places, the logarithm of 2.48 equals 0.394. The log of 100 is 2. The log of 248 is therefore 0.394+2. In logarithmic terms:

$$248 = 2.48 \times 100$$

$$\log 248 = \log(2.48 \times 100)$$

$$= \log 2.48 + \log 100$$

$$\log 2.48 = 0.394 \qquad \log 100 = 2$$

$$\log 248 = 0.394 + 2 = 2.394$$

The logs and antilogs can be determined by either the use of a log table or a calculator. The following example shows how separating the integers and decimals (or characteristics and mantissas) of the logs simplifies multiplication and division.

Example 11.1:   Using the log table, calculate $342 + 0.0239$

Solution: The first step is to convert the numbers into exponential notation.

$$342 = 3.42 \times 10^2$$

$$0.0239 = 2.39 \times 10^{-2}$$

The next step is to find the logs of the numbers. From the tables in Appendix A, log 3.42 = 0.534 and log 2.39 = 0.378.

$$\log(3.42 \times 10^2) = (\log 3.42) + 2 = 0.534 + 2$$
$$\log(2.39 \times 10^{-2}) = (\log 2.39) - 2 = 0.378 - 2$$

Since this problem involves division, the logs are subtracted.

$$0.534 + 3$$

$$-\quad \underline{0.378 - 2}$$

$$0.156 + 5$$

We now look up the antilog of 0.156 in the tables. The antilog of 0.156 is 1.43. The antilog of 0.156 + 5 is therefore $1.43 \times 10^5 = 14{,}300$.

Given that the logs were accurate to three significant figures, the answer will only be accurate to three significant figures. The answer is therefore approximate.

11.1.1 Multiplication of Exponents

Logarithms may also be used to determine the value of exponentials of large or small numbers. Another exponential equation learned in Chapter 3 was:

$$\left(a^m\right)^n$$

Using logarithmic terms, the formula may be written as:

$$\log_a M^n = n \log_a M \qquad\qquad (11\text{-}3)$$

From Table 11.1 we see that $\log 2 = 0.3$. Using Eq. 11-3, we can calculate the value of the 2^3 by multiplying the log of 2 by 3 and then taking the antilog.

$$\log 2 = 0.30$$

$$\log 2^3 = 3 \log 2 = 3(0.30) = 0.90$$

$$\text{antilog} 0.90 = 8$$

An observant student might ask, "If calculators and computers are available, why do we need to learn how to use logarithms for multiplication and division?" Without question, calculators and computers are faster and more accurate than log tables. However, as noted earlier, calculators and computers use logarithms to multiply and divide. Companies and countries that can design and build calculators and computers enjoy great economic advantages. For this reason, students are still taught how logarithms are used to do arithmetic calculations.

Exercise Set 11.1

Use Appendix A to find the common logs of the following numbers:

1.	3.57	3.	4.09	5.	999
2.	1.64	4.	72.1	6.	1,260

Use Appendix A to find the common antilogs for the following numbers:

7. 0.467   9. 0.894   11. 3.961

8. 0.705   10. 1.344   12. 5.785

Use common logs to compute the following operations:

13. 1,000 x 35.7   15. 13 x 49   17. $\left(4.7 \times 10^{-3}\right)\left(3 \times 10^{8}\right)$

14. $\dfrac{1}{8.62}$ 16. $\dfrac{5.5}{1.3}$ 18. $6.02 \times 10^{23} \div 3.47$

11.2 NATURAL LOGARITHMS

As discussed earlier, logs to the base 10 were developed for a particular reason. Common logarithms were developed because they make it easy to do complex arithmetic problems. Another type of logarithm, *natural logarithms*, were developed for a different reason. The base of natural logs is a special number known as *e*. The number *e* is an irrational constant like π. Its value is approximately 2.73. The term *natural* logarithm comes from the fact that *e* occurs in many areas in nature. The number *e* and natural logs occur in physics, business, biology, and many other areas. To fully appreciate the meaning of *e*, you need to take a course in calculus. This discussion merely offers an introduction to the concept of natural logarithms.

Natural logs are denoted by the term *ln*. All of the properties discussed for common logarithms also apply to natural logs. For example

$$\ln e^x = x \qquad\qquad (11\text{-}4)$$

In Chapter 6, we looked at linear relationships. Another type of mathematical relationship which is common in business and science is *exponential relationships*. The following discussion helps illustrate the difference between linear and exponential relationships. Understanding the difference between the two types of relationships is a key factor in understanding *e*.

<u>Example 11.2:</u> Cheryl and Ralph run a laundromat business. Their business does very well and they regularly try to expand their business by adding more washing machines and dryers. For every machine they add, they can expect a regular amount of increased income. For their first few years of business they have a policy of adding four new machines every year. Since they add a constant number of machines every year and the machines are well used, their income increases at a regular rate every year. If you graphed their income over time, it would look something like this:

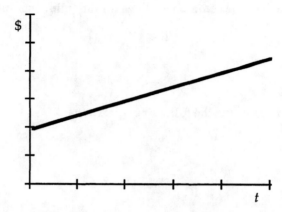

Figure 11.1 Linear Relationship

After being in business a few years, Cheryl and Ralph decide on a new policy. Rather than adding a constant number of machines every year, they're going to use half of their income each year to purchase new machines. With each new machine, their income increases; and as their income increases every year, the number of new machines they purchase increases every year. If you graphed their income over time, it would look something like this:

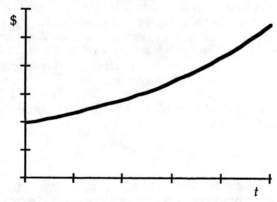

Figure 11.2 Exponential Relationship

Under the new policy, the relationship between their income and time is known as an exponential relationship. Exponential relationships are common in business. For example, if you left money in a savings account and neither deposited nor withdrew money, interest would be added to the account and the balance would grow. The balance of the account would grow exponentially.

Exponential growth is also common in biology. For example when a new type of plant or animal enters a new habitat and is successful at surviving and reproducing, its increase in population every year will depend on the size of the population that year. Each year's growth will be greater than the previous year's.

There are also examples of exponential decreases in nature. Radioactive materials such as plutonium *decay* exponentially over time. As they decay, they become less harmful. Natural logarithms are used to determine how long it will take radioactive waste from nuclear power plants to decay to a point where they'll no longer create health risks such as causing cancer.

Using logarithms to understand and predict the exponential decay of radioactive materials serves another important function in science. A particular form of the element carbon, C^{14} (carbon 14), decays exponentially into the more common form of carbon, C^{12} (carbon 12). Since the amount of C^{14} remains relatively constant in the atmosphere, the amount of C^{14} in plants and animals that absorb it stays relatively constant. When the plant or animal dies and stops absorbing C^{14}, the C^{14} begins to decay. If you find a fossil of the plant or animal, you can measure the amounts of C^{14} and other forms of carbon in it. By comparing the amount of C^{14} with the other forms of carbon, you can determine how much C^{14} has decayed. Since scientists have determined the rate of decay of C^{14}, you can tell how long the C^{14} in the fossil has been decaying. In other words, you can tell how old the fossil is. This is how scientists determine the age of fossils and archeological artifacts.

All of the examples discussed here involve natural logarithms. Examples of the exponential growth of a plant or animal population can usually be estimated or *modeled* with equations of the form:

$$a = a_0 e^{kt}$$ where a = the population size at time t (11-5)
a_0 = the initial population size
k = a constant specific to the growth.
t = time since plant or animal entered new environment

Exponential decay can be modeled with equations of the form:

$$a = a_0 e^{-kt}$$ where a = the amount of material at time t (11-6)
a_0 = the initial amount of material
k = a constant specific to the decay
t = time since the initial amount of material was created

Example 11.4: A bio-technology firm has genetically engineered a new type of insect to control weeds. The insect attacks a certain type of weed and is harmless to all other types of plants. Before selling the insect to farmers, the firm wants to understand some of the insect's growth characteristics. It conducts a test to determine how fast the insect will multiply with an unlimited supply of food. It releases 121 insects into an enclosed area with food and then counts the insect population at different times. At 50 days the population is counted to be 537. Determine:

a.) The equation that estimates the population of the insects with an infinite food supply as a function of time in days.

b.) The size of the population at 80 days.

Solution: a.) Since population growth is exponential, the growth of the insect population can be modeled by Equation 11-5:

$$a = a_0 e^{kt}$$ where a = the population size at time t
t = time, in days
a_0 = the initial population size = 121
k = a constant

The problem gives us the value of all of the variables except k. To determine the value of k, we substitute all of the known values into the equation and solve for k. At $t = 50$ days

$$a = a_0 e^{kt}$$

$$537 = 121 e^{k(50)}$$

$$e^{50k} = \frac{537}{121}$$

$$\ln e^{50k} = \ln \frac{537}{121}$$

$$50k = \ln \frac{537}{121}$$

$$k = \frac{\ln \dfrac{537}{121}}{50}$$

$$= 0.030$$

We can then substitute the value of k into the original equation.

The growth of the population of the insects, a, with an infinite supply of food can be modeled with the equation:

$$a = 121 e^{0.03t} \quad \text{where } t = \text{time, in days}$$

b.) To calculate the population at 80 days, we substitute 80 for t in the equation. At 80 days, the population could be expected to have a size of:

$$a = 121 e^{0.03(80)}$$

$$\approx 1,313$$

At 80 days, the population of the insects could be expected to reach 1,313.

Exercise Set 11.2

1. A bio-technology firm is developing a new strain of bacteria to convert milk to cheese more quickly. To learn the growth characteristics of the bacteria, the firm releases 112 of the organism into an environment with an infinite food supply. The population of the bacteria from this point will be given by the equation:

$$p = 112e^{kt} \quad \text{where} \quad \begin{aligned} p &= \text{the population size at time } t \\ t &= \text{time, in hours} \\ k &= \text{the growth rate constant} \end{aligned}$$

At 30 hours the researchers count 356 organisms. What is the growth rate constant, k, of the bacteria and how many bacteria could be expected at 40 hours after the organisms were first released?

2. A bone fragment is found in the ruins of an ancient Aztec city and is tested for carbon dating. The decay of C^{14} is given by the equation:

$$p = p_o e^{-kt} \quad \text{where} \quad \begin{aligned} p &= \text{the population size at time } t \\ t &= \text{time, in years} \\ k &= \text{the growth rate constant} \end{aligned}$$

The C^{14} reading indicates that 73% of the original C^{14} remains. In other words, $p/p_o = 0.73$. The decay constant, k, of C^{14} is 0.000126. How old is the fragment?

3. An accident releases radioactive cobalt into the surrounding area. It is estimated that the radiation level must decay to 5% of the present level (i.e., $p/p_o = 0.05$) for the area to become safe again.

a.) If the decay constant for cobalt is 0.135, how many years will it before the area is safe to inhabit again?

b.) If the accident had released uranium 235 (U_{235}) with a decay constant of 9.76 x 10^{-10} instead of cobalt, how long would it take to decay to 5% of the original amount?

REFERENCES

C. Edwards, Jr., D. Penney, *Elementary Differential Equations with Applications*, Prentice-Hall, Englewood Cliffs, New Jersey, 1985. Formulas for exponential growth and decay

Chapter 12

CYCLICAL MOTION

In Chapter 6 we learned to understand linear motion by breaking it down into components of distance, speed, and time. In Chapter 9, we learned how to understand parabolic motion by breaking it down into horizontal velocity, vertical velocity, and vertical acceleration. Another type of motion which occurs often in sports is cyclical motion.

When dribbling down the court, a basketball player moves her hand in sync with the motion of the ball. To develop quickness, a boxer punches a hanging bag in sync with the motion of the bag. Hurdling requires a cyclical motion as does running itself and swimming. An athlete jumping rope moves the rope with his hands and jumps in a particular rhythm to miss the rope. School children who play double dutch jump to an even more complex rhythm.

Cyclical motion occurs in countless other areas in the physical world. The pistons in an engine will fire in a precise sequence thousands of times in a minute. A mechanic will check the engine using a flashing timing light to make sure it's running properly. Electrical motors spin running everything from pumps to power tools. A radio station will broadcast radio waves which, when hitting an antenna attached to a tuned-in stereo, will vibrate a speaker creating music. Similarly, a television receives electromagnetic waves which cause a stream of light to hit thousands of dots on a screen millions of times per second to create a picture.

Like many of the examples in sports, the operation of machinery involves the motion of two or more objects which must move perfectly together in order to work. The basketball player's hand must move perfectly with the ball and the pistons in an engine must fire perfectly in order to turn an axle. To make things move together properly requires a thorough understanding of cyclical motion. The same math tools are used for all types of cyclical motion from that of a motor to the cycles of a TV or computer screen.

As one might expect, the math involved in cyclical motion is based on circular motion. Once the mathematics of circular motion is understood, it can be expanded to understand virtually any other type of cyclical motion.

To examine circular motion, let's take a closer look at an athlete skipping rope. To monitor the motion of the rope, we'll place an xy plane on the athlete and put his hands at the origin. The ends of the rope (his hands) remain in a relatively steady position at the origin. The tip of the rope extends 1 meter in the plane of the graph. The rope remains relatively straight as it rotates. The rope rotates counterclockwise and just barely touches the ground.

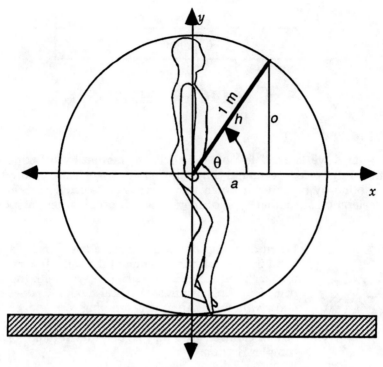

Figure 12.1   Jumping Rope

Skipping rope requires keeping track of the location of the tip of the rope. The athlete
must jump when the tip of the rope is near his feet. The position of the tip can be
described by either:

    1. The angle of the rope, θ, or

    2. The height (y-value) of the tip above or below the x-axis.

In the Figure 12.1, the rope has rotated to an angle θ. A triangle is created which helps
keep track of the position of the tip of the rope. The leg of the triangle which represents
the height of the tip above the x-axis is marked with an o, for the length of the leg *opposite*
the angle. The other leg is marked with an a for the length of the leg *adjacent* to the angle.
The hypotenuse of the triangle is marked with an h. Analyzing cyclical motion involves
understanding the relationships that exist between θ and the lengths of the sides of the
triangle. For a right triangle of a certain shape, the relationships between θand the
lengths of the sides remain constant regardless of its size. Because of the triangle that is
formed in the figure, the branch of mathematics which studies the relationship between
the angles and sides is known as *trigonometry*.

12.1 THE TRIGONOMETRIC FUNCTIONS

There are six trigonometric functions which are used to describe the relationships
between θ and the sides of a right triangle. The trigonometric functions have numerous
uses other than to analyze cyclical motion. One of their major uses is to calculate

unknown distances. In fact, the use of trigonometry to calculate distances forms the basis for analyzing cyclical motion.

The Pythagorean Theorem provided one method of finding the unknown length of a side of a right triangle when the lengths of the other two sides were known. By using the trigonometric functions, we can determine the lengths of the other two sides by knowing the length of one of the sides and the size of one of the angles. The following section discusses how each of the six trigonometric functions can be used to calculate distances.

12.1.1 The Sine Function

The most commonly used relationship in trigonometry is the one between θ and the ratio of the opposite leg to the hypotenuse. This relationship is known as the *sine* of θ. This relationship is expressed sinθ. The relationship is given by the following equation:

$$\sin \theta = \frac{\text{length of the opposite leg}}{\text{length of the hypotenuese}} = \frac{o}{h} \qquad (12\text{-}1)$$

In the jump rope example, the length of 1 meter to the tip of the rope was chosen for a particular reason. Since the discussion states that the tip of the rope remains 1 meter from the athlete's hands, the length of the hypotenuse is always 1. Recall that any real number divided by one equals the number. If h is always equal to 1, the sine of the angle will always be equal to the length of the opposite leg. In mathematical terms:

$$\text{for } h = 1, \qquad \frac{o}{h} = \frac{o}{1} = o$$

In trigonometry, the circle with a radius of 1 is known as the *unit circle*. The unit circle is used to explain all of the trigonometric relationships. Figure 12.2 shows the unit circle on the left. The gray lines represent the rotating hypotenuse. The dotted lines represent the fluctuating lengths of o. The height of o is plotted on the right as function of θ. In other words, the locus is the curve $y = \sin\theta$.

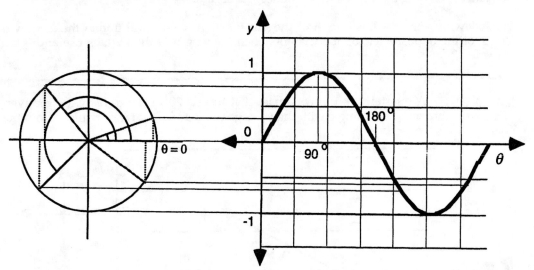

Figure 12.2 The Sine Function

The curve that is created by plotting the height of the opposite leg is known as a *sine wave*. Note that when θ= 0° in the unit circle, an infinitely thin triangle is formed. In other words when θ = 0°, the length of the opposite leg *o* equals 0. Therefore, sin0°= 0. When θ = 90° (π/2 radians), the length of *o* equals the length of *h*. Another infinitely thin triangle is formed. The sine of 90° equals h/h or 1. At 180° (π radians), the length of the opposite leg is back down to 0. Therefore sin180° also equals 0. After θ passes 180°, *o* becomes negative. Therefore sinθ becomes negative after 180°. At 270° (3π/2 radians), the length of *o* is in the negative direction. The sine of 270° equals -h/h or -1. Since the length of the opposite leg can only get as long as the hypotenuse, 1, the sine function fluctuates between 1 and -1. When the rope reaches 360°, *a* is again back down to zero. The sine of 360° therefore equals 0. After θ passes 360°, θ and sinθ again become positive. The sine of an angle can also be taken when θ is negative. By looking at the unit circle, one can see that the angle of 270° is the same as -90°.

Ordinarily, when trigonometric functions are plotted on graphs, the angle of rotation is given in terms of radians. The graph of a sine wave is shown in Figure 12.3.

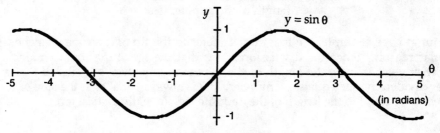

Figure 12.3 Sine Wave

The values of sinθ and the other trigonometric functions for different angles are given in Appendix B. Tables which give the values of the different functions are commonly known as *trig tables*. The values are given for angles in degrees or radians. Calculators are often programmed with the trigonometric functions. If you use a calculator to determine the value of a trigonometric function, you must be conscious of whether the calculator is reading the angle in terms of degrees or radians.

The following example illustrates how the sine function can be used to find the unknown length of the opposite leg, *o*, of a triangle when the length of the hypotenuse and θ are known.

Example 12.1: A quarterback throws a pass that travels 20 yards to his wide receiver. The pass is thrown at a 30° angle with the yardage markers. How far downfield from the quarterback is the receiver?

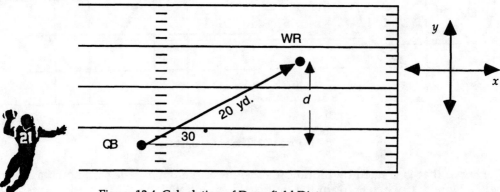

Figure 12.4 Calculation of Downfield Distance

Solution: We notice that the distance of the pass represents the hypotenuse of a triangle. The distance downfield, d, represents the length of the opposite leg of this triangle. From Eq. 12-1:

$$\sin \theta = \frac{o}{h}$$

Therefore $\sin 30° = \dfrac{d}{20}$ where d = the distance downfield from

the quarterback

Solving the equation for d, $d = 20\sin 30°$

By using a calculator or by finding the value in Appendix B:

$$d = 20(0.5000)$$

$$= 10 \text{ yards}$$

The pass travels 10 yards downfield.

12.1.2 The Cosine

The sine of the angle measures the ratio of the length of the opposite leg to the length of the hypotenuse. It is sometimes useful to measure the ratio of the length of the *adjacent leg* to the hypotenuse. The ratio of the length of the adjacent leg, a, to the length of the hypotenuse is known as the *cosine* of the angle. The relationship is given by the equation:

$$\cos \theta = \frac{\text{length of adjacent leg}}{\text{length of hypotenuese}} = \frac{a}{h} \qquad (12\text{-}2)$$

If we rotated the unit circle in Figure 12.2 90° to the left, we would be able to plot the changing length of the adjacent leg. The length of the adjacent leg fluctuates exactly like the length of the opposite leg. The graph of the cosine function is almost identical to the graph of the sine function. The only difference is that the crest ($y = 1$) of the sine function occurs at 90° ($\pi/2$ or 1.57 radians). The crest of the cosine function is at 0°. The graph of the cosine compared with the sine function is shown in Figure 12.5.

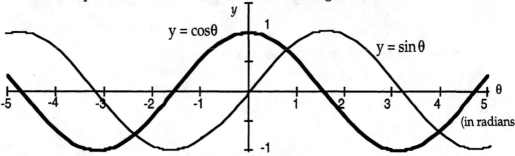

Figure 12.5 Cosine and Sine Functions

As you can see from the figure, the cosine function, is off 90° ($\pi/2$ radians) from that of the sine function. Given that the graph of the cosine function is 90° off from the sine function, we can see one of the first relationships between trigonometric functions.

$$\cos\theta = \sin(\theta + \frac{\pi}{2})$$ (12-3)

When functions have graphs like the sine and cosine functions, they are said to be *sinusoidal*.

The cosine function allows us to find the length of the adjacent leg if we know the length of the hypotenuse and θ. In terms of the football pass problem, it allows us to find the lateral (x-distance) length of the pass.

Example 12.2: In the 20-yard pass of Example 12.1, determine the lateral distance (x-distance) of the pass.

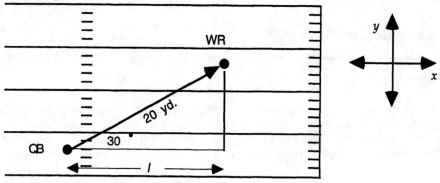

Figure 12.6 Calculation of Lateral Distance

Solution: The lateral distance of the pass represents the length of the adjacent leg, l, of the triangle.   Using Eq. 12-2:

$$\cos\theta = \frac{a}{h}$$

Therefore $\cos 30° = \dfrac{l}{20}$ where l = the lateral distance from

the quarterback, in yards

Solving the equation for l, $l = 20\cos 30°$

By using a calculator or by finding the value in Appendix B:

$$= 20(0.8660)$$

$$\approx 17 \text{ yards}$$

The pass travels approximately 17 yards laterally.

Another important use of the sine and cosine functions is that they allow us to easily determine the vector of a motion if we know the angle and magnitude of the motion. Using the process shown in the last two problems, we can see that the displacement vector of the pass is $17i + 10j$. Any resultant vector R with a magnitude of R and having an angle θ with the x-axis can be given by the equation:

$$R = R\cos\theta\, i + R\sin\theta\, j \qquad\qquad (12\text{-}4)$$

12.1.3 The Tangent Function

It is often useful to have trigonometric functions for the ratios of the two legs of a right triangle. These functions are known as the *tangent* and *cotangent*. The tangent function is defined as follows:

$$\tan \theta = \frac{\text{length of opposite leg}}{\text{length of adjacent leg}} = \frac{o}{a} \qquad\qquad (12\text{-}5)$$

A line which touches only one point on a circle (like touching the outside of the circle) is said to be *tangent* to the circle. This is where the tangent function gets its name.

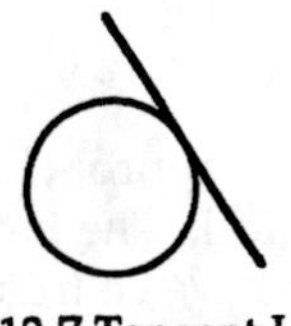

The tangent and cotangent functions can also be defined with the unit circle. With these functions, it is the length of one of the legs which remains at 1 rather than the length of the hypotenuse. As

Fig. 12.7 Tangent Line

can be seen in Figure 12.8, the lengths of the opposite leg and hypotenuse can become very large. Unlike the sine function, the length of the opposite leg can be many times the length of the adjacent leg. The tangent of θ can therefore be much greater than 1. The tangent function is graphed in Figure 12.8.

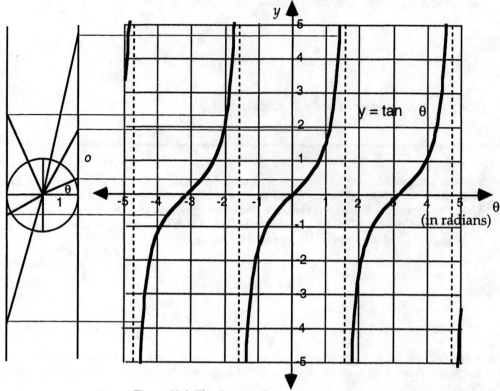

Figure 12.8 The Tangent Function

From the figure, we can see that when θ = 0°, the length of the opposite leg is 0. An infinitely thin triangle is formed. Therefore:

$$\tan 0 = \frac{0}{1} = 0$$

As θ approaches 90° (π/2 radians), the hypotenuse of the triangle becomes almost vertical. The tangent of θ becomes very large. As θ reaches 90°, the leg of the hypotenuse will have an infinite length. Therefore:

$$\tan\frac{\pi}{2} = \frac{\infty}{1} = \infty$$

As θ passes 90°, the length of the adjacent leg becomes negative. Therefore tanθ becomes negative. As θreaches 180° (π radians), the length of the opposite leg is again 0. Therefore:

$$\tan \pi = 0$$

As θ reaches 270° (3π/2 radians), the hypotenuse of the triangle again becomes vertical but in the negative direction. The tangent of θ will then have a value of negative infinity. In mathematical terms:

$$\tan\frac{3\pi}{2} = \frac{\infty}{-1} = -\infty$$

Like the sine and cosine functions, the tangent function is also used to calculate distances. The following problem provides an example of how the tangent function can be used to calculate unknown distances.

Example 12. 3: During a softball game, a second baseman fields a grounder right on the baseline between first and second. If the ball is hit at a 37° angle with the first base line, how far is the throw to first base?

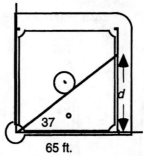

Figure 12.9 Distance to First Base

Solution: This problem gives us the values of θ and the length of the adjacent leg and asks for the length of the opposite leg. The length of the opposite leg can be found with the tangent function. Using Eq. 12-5:

$$\tan 37° = \frac{d}{65}$$

$$d = 65 \tan 37°$$

The value of the tangent of 37° can be found either with a calculator or from the trig tables in Appendix B. To four digits, the value is 0.7536. Therefore:

$$d = 65(0.7536)$$

$$\approx 49$$

From the point at which the grounder is fielded, the throw to first base is approximately 49 ft.

12.1.4 The Cotangent Function

The cotangent function is defined as follows:

$$\cot \theta = \frac{\text{length of adjacent leg}}{\text{length of opposite leg}} = \frac{a}{o} \qquad (12\text{-}6)$$

The cotangent function can also be defined with the unit circle. With the cotangent, it is the length of the opposite leg which remains constant at 1. Similar to the relationship between the sine and the cosine, the graph of the cotangent can be visualized by rotating the unit circle of the tangent function 90°.

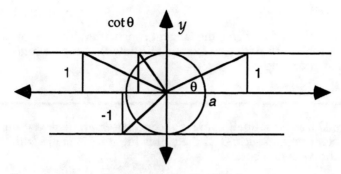

Figure 12.10 Cotangent Unit Circle

The cotangent is graphed as shown in Figure 12.11.

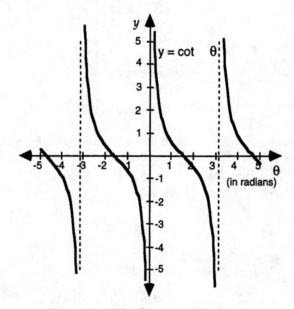

Figure 12.11 The Cotangent Function

Note that the cotangent function is the reciprocal of the tangent function. In mathematical terms:

$$\cot \theta = \frac{1}{\tan \theta}$$

(12-7)

Calculators are usually not programmed with the cotangent function. To determine the cotangent of an angle, you can either use a trig table or use a calculator to determine the tangent of the angle and then calculate the reciprocal of that value.

Example 12.4: A shortstop fields a grounder on the baseline between second and third. If the ball is hit at a 62° angle with the first base line, how far is the throw to third base?

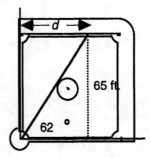

Figure 12.12 Distance to Third Base

Solution: This problem gives us the values of θ and the length of the opposite leg and asks for the length of the adjacent leg. Note that the distance of the throw to third base equals the length of the adjacent leg of the triangle. The length of the adjacent leg can be found with the cotangent function. Using Eq. 12-6:

$$\cot 62° = \frac{d}{65}$$

$$d = 65 \cot 62°$$

The value of the cotangent of 62° can be found in Appendix B. To four digits, the value is 0.5317. Therefore:

$$d = 65(0.5317)$$

$$\approx 34.6$$

If trig tables are not available to provide the cotangent value, one can determine the solution by dividing by the tangent of the angle.

$$\tan 62° = \frac{65}{d}$$

$$d = \frac{65}{\tan 62°}$$

$$= \frac{65}{1.881}$$

$$= 34.6$$

From the point at which the grounder is fielded, the throw to third is approximately 34.6 ft.

12.1.5 The Secant Function

Two other functions complete the relationships which are used in trigonometry. With the *secant* and the *cosecant* functions, the length of the hypotenuse is in the numerator. The secant function is defined as follows:

$$\sec \theta = \frac{\text{length of hypotenuese}}{\text{length of adjacent leg}} = \frac{h}{a} \tag{12-8}$$

A secant is a line that passes through a circle. Similar to the tangent function, the secant function gets its name from the fact that it represents the changing length of a secant line which, in the case of the unit circle, is the hypotenuse of a triangle. The unit circle used for the tangent function in Fig. 12.8 could also be used to illustrate the secant function. In this case, it would be the length of the hypotenuse that would be plotted as a function of θ. Like the tangent function, the secant function also fluctuates between infinity and negative infinity. The secant is graphed as shown in Figure 12.14.

Fig. 12.13 Secant Line

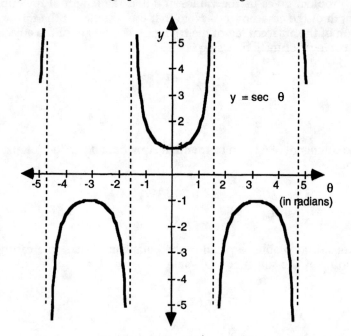

Figure 12.14 The Secant Function

Note that:

$$\sec \theta = \frac{1}{\cos\theta} \tag{12-9}$$

Like the cotangent function, the secant function is usually not programmed on calculators. If trig tables are not available, one can divide by the cosine of the angle rather than multiply by the secant of the angle.

Example 12.5: A second baseman fields a grounder right on the baseline between first and second. If the ball is hit at a 23° angle with the first base line, how far is the throw to home?

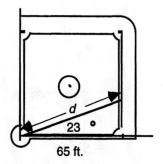

Figure 12.15 Distance to Home

<u>Solution:</u> This problem gives us the values of θ and the length of the adjacent leg and asks for the length of the hypotenuse. The length of the hypotenuse can be found with the secant function. Using Eq. 12-8:

$$\sec 23° = \frac{d}{65}$$

$$d = 65 \sec 23°$$

The value of the secant of 23° can be found in Appendix B. To four decimal places, the value is 1.0864. Therefore:

$$d = 65(1.0864)$$

$$\approx 70.6$$

If trig tables are not available to provide the secant of the angle, one can determine the value by dividing by the cosine of the angle.

$$\cos 23° = \frac{65}{d}$$

$$d = \frac{65}{\cos 23°}$$

$$= \frac{65}{0.9205}$$

$$\approx 70.6$$

From the point at which the grounder is fielded, the throw to home is approximately 70.6 ft.

12.1.6 The Cosecant Function

The cosecant function represents the ratio of the length of the hypotenuse to the length of the opposite leg. In mathematical terms:

$$\csc \theta = \frac{\text{length of hypotenuese}}{\text{length of opposite leg}} = \frac{h}{a} \qquad (12\text{-}10)$$

The cosecant is graphed as shown in Figure 12.16.

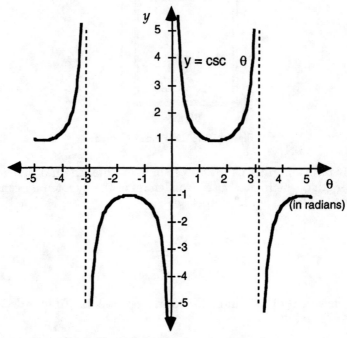

Figure 12.16 The Cosecant Function

Note that the cosecant is the inverse of the sine function:

$$\csc \theta = \frac{1}{\sin\theta}$$

(12-10)

Like the cotangent and secant functions, calculators are usually not programmed with the cosecant function. The cosecant of an angle can be found by determining the reciprocal of the sine of the angle.

Example 12.6: A third baseman fields a grounder right on the baseline between second and third. If the ball is hit at a 79° angle with the first base line, how far is the throw to home?

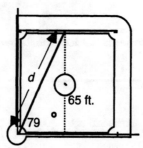

Figure 12.17 Throw to Home

Solution: This problem gives us the values of θ and the length of the opposite leg and asks for the length of the hypotenuse. The length of the hypotenuse can be found with the cosecant function. Using Eq. 12-10:

$$\csc 79° = \frac{d}{65}$$

$$d = 65 \csc 79°$$

The value of the cosecant of 79° can be found in Appendix B. To four decimal places, the value is 1.0187. Therefore:

$$d = 65(1.0187)$$

$$\approx 66.2$$

If trig tables are not available to provide the secant value, one can determine the solution dividing by the sine of the angle.

$$\sin 79° = \frac{65}{d}$$

$$d = \frac{65}{\sin 79°}$$

$$= \frac{65}{0.9816}$$

$$= 66.2$$

From the point at which the grounder is fielded, the throw to home is approximately 66.2 ft.

Perhaps the field which uses the trigonometric functions to calculate unknown distances the most is surveying. By using these functions, surveyors can calculate distances that may be impossible to measure directly such as across rivers, through forests, and under mountains. Similarly, when machinery is designed, the trigonometric functions may be used to calculate the lengths of parts or the distances between objects.

Exercise Set 12.1

1. On an option play, a quarterback laterals 8 yards to his halfback at a -20° angle with the yardage markers.

a.) How far backfield from the quarterback is the halfback when he receives it?
b.) What is the lateral distance (x-distance) between them?
c.) What displacement vector represents the lateral?

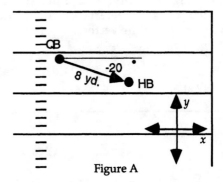

Figure A

2. A second baseman fields a grounder right on the baseline between second and third. The ball is hit at a 55° angle with the first base line. (65 feet between bases)
a.) How far does the ball travel to the infielder?
b.) How far is the throw to third base?

3. A backboard is suspended by a cable as shown in Figure B
a.) Determine the length of the moment arm, l, of the cable.
b.) Determine the distance between the points on the wall where the cable and poles are fixed.
c.) What is the length of the cable?

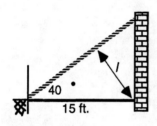

Figure B

4. From a distance, surveyor watches the space shuttle liftoff from Cape Canaveral. She knows she is 5 miles from the launch pad. A few seconds after liftoff, she notices that the shuttle is 32° above the horizon.
a.) At this instant, what is the altitude of the shuttle in feet?
b.) What is her line of sight distance from the shuttle?

The following problems give the magnitudes of some resultant vectors and their angles relative to the positive side of the x-axis. Determine the i and j components of the vectors.

5. 46 at 12°

6. 112 at 67°

7. 0.04 at 7°

8. 67 at 111°

9. 14 at -31°

10. 513 at 242°

11. 300 at 3.5605 radians

12. 79 at 5.5501 radians

13. 100 at 0.0349 radians

14. 120 at 1.5533 radians

The discussion of parabolic motion in Chapter 9 gave several formulas in terms of v_{0x} and v_{0y} where $v_{0x}=$ the initial velocity of the object in the x-direction; and $v_{0y}=$ the initial velocity of the object in the y-direction. For each of the following equations, replace the v_{0x} and v_{0y} terms with v, the magnitude of the resultant, and the sine or cosine of θ.

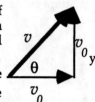

15. $x = v_{0x}t$

16. $v_x = v_{0x}$

17. $y = -\dfrac{gt^2}{2} + v_{0y}t$

18. $v_y = v_{0y} - gt$

19. $y = -\dfrac{g}{2v^2_{0x}}x^2 + \left(\dfrac{v_{0y}}{v_{0x}}\right)x$

20. $H = \dfrac{v^2_{0y}}{2g}$

21. $R = \dfrac{2v_{0x}v_{0y}}{g}$

12.2 THE ARC FUNCTIONS

The trigonometric functions tell us the ratios of the lengths of the sides of a right triangle if we know the value of θ and the length of one side. It is often desirable to work backwards and determine the angle if we know the lengths of the sides. The *arc* functions allow us to do this. With the arc functions, the relationships of θ and the ratios are reversed. In mathematical terms:

$$\text{for } \sin\theta = \frac{o}{h} \qquad \arcsin\frac{o}{h} = \theta$$

(12- 12)

The arc functions are often abbreviated with a -1 exponent. For example:

$$\arctan x = \tan^{-1}x$$

Calculators usually determine the arc functions with an inverse key. For example, rather than just hitting the "sin" key to calculate the sine of an angle, one would key in the ratio, hit the "INV" key and then the "sin" key to determine the angle. Again, one must be conscious of whether the calculator is in a degrees or radians mode.

Example 12.7: A quarterback throws an 18-yard pass to his wide receiver. The receiver catches the ball 12 yards downfield from the quarterback. At what angle with the yard markers was the pass thrown?

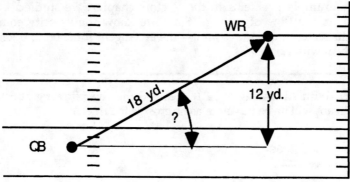

Figure 12.18 Calculation of Pass Angle

Solution: The sine of the angle would be given by the equation:

$$\sin\theta = \frac{12}{18} \qquad \text{where } \theta = \text{the angle between the pass}$$

and the yard markers

Given the relationship of Equation 12-11:

$$\theta = \sin^{-1}\frac{12}{18}$$

$$= \sin^{-1}(0.6667)$$

By using a calculator or from Table B in the Appendix:

$$= \sin^{-1}(0.6667)$$

$$= 42°$$

The pass travels at a 42° angle with the yard markers.

The *arctangent* function serves a special purpose with vectors. The arctangent function allows us to determine the angle of a vector directly from its *i*- and *j*-components. Similar to the arcsin function, the arctangent is abbreviated as "arctan." In mathematical terms:

$$\text{for} \quad \tan \theta = \frac{o}{a}$$

(12-13)

$$\arctan \frac{o}{a} = \tan^{-1} \frac{o}{a} = \theta$$

In one of the example problems in the Vectors chapter, we studied a person rowing perpendicular to the flow of a river. Since we know the velocity components of the rowing and the water current, we can use the arctangent function to determine the angle at which the boat will travel.

Example 12.8: With Al rowing at 12 ft./sec and the river flowing at 5 ft./sec., at what angle downstream will he be traveling as he crosses the river?

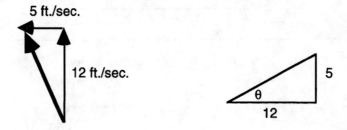

Figure 12.19 Calculation of Angle of Boat Travel

Solution: The resultant velocity vector is -5*i* + 12*j*. The angle is determined by taking the arctan of $\frac{5}{12}$.

$$\tan^{-1} \frac{5}{12} = \tan^{-1} 0.41\overline{6} = 23°$$

By using a calculator or looking up 0.4167 in the trig tables, we see that θ equals 23°. This means that the boat will veer 23° downstream.

Exercise Set 12.2

1. A football team attempts a field goal from the position shown in Figure A. If the kick is long enough and there is no wind, what is the range of angles, θ, that will result in a field goal.

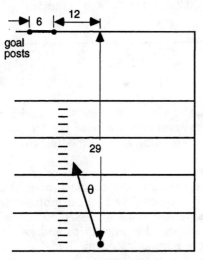

Figure A

2. A racquetball strikes a vertical wall with a velocity vector of $3i - 8j$. Because of sidespin and friction on the wall, the ball departs the wall with a velocity vector of $4i + 6j$.
a.) What was the angle of incidence, θ_1?
b.) What was the angle of rebound, θ_2?

Figure B

3. A first baseman and a second baseman are positioned on the baseline as shown in the figure. At what angle with the firstbase line will a grounder pass directly between them?

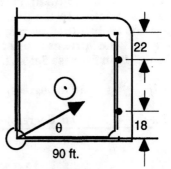

Figure C

4. A surveying team is setting the stakes for the foundation of a house. The instrument which is used to measure angles (known as a *theodolite*) rests on a tripod and must be perfectly level to be accurate. It is therefore somewhat time consuming to set up. As a result, it is preferable to set it up at one location and take all of your measurements from that point. The dimensions of the foundation are shown in Figure D.

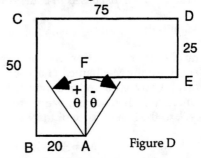

Figure D

All of the foundation's corners are right angles. All angles and distances will be measured from point A. To expedite the process, a table will be prepared ahead of time which contains the angles and distances from point A. All angles will be in degrees relative to the line AF. Given the dimensions, complete Table A.

Point	θ (degrees)	Distance (ft.)
B		
C		
D		
E		
F		

Table A

For the following vectors, determine the magnitudes their resultant vectors and their angles relative to the positive side of the x-axis. In other words, use the reverse of the process used to do problems 5-14 in Exercise Set 12.1.

5. $2i + 7j$ (θ in degrees)

6. $100i + 0j$ (θ in degrees)

7. $-14i - 60j$ (θ in degrees)

8. $64i - 6j$ (θ in radians)

9. $-36i + 72j$ (θ in radians)

10. $0.23i + 0.18j$ (θ in radians)

12.3 SINUSOIDAL FUNCTIONS

In the preceding sections, we learned how the trigonometric functions can be used to calculate distances. By looking at their graphs, one can see how the trigonometric functions are cyclical. For all of the functions, each rotation around the unit circle creates a new cycle.

Recall that of the trigonometric functions, only the sine and cosine functions do not alternate between negative and positive infinity. Since few cyclical motions involve negative or positive infinity, the sine and cosine functions are used to model most cyclical motions. This section discusses some sinusoidal relationships and two of their key parameters, *amplitude* and *frequency*. More sophisticated methods of modeling cyclical motions are discussed in courses teaching Fourier analysis.

12.3.1 Amplitude

By using the cosine function we can expand on the one-arm curl problem discussed in the Weight Room Mechanics chapter. A close look shows that there is a sinusoidal relationship between θ and the moment that is created by the weight. While calculating the moment which resulted, we were limited by the need to know the perpendicular distance, l, to the line of force of the weight, w. Recall that:

$$M = Fl \quad \text{where} \quad M = \text{the moment}$$
$$F = \text{the force, and}$$
$$l = \text{the length of the moment arm.}$$

With the cosine function, we can calculate the perpendicular distance, and therefore the moment, for any angle θ. In this example, the length of the moment arm, l, equals the length of the adjacent leg, a.

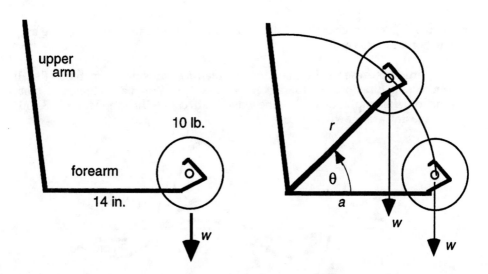

Figure 12.20 Calculation of Maximum Moment

From the diagram, we can see that:

$$\cos\theta = \frac{a}{r}$$

$$a = r\cos\theta$$

The moment M equals Fl, where $l = a = 14$ in. and $F = w = 10$ lb. Therefore the moment at any angle θ is given by the equation:

$$M = Fl = (14)(10) = 140\cos\theta, \text{ with units in in.·lb.}$$

An engineer who comes across the same configuration in a piece of machinery may need to know the exact moment for every value of θ. For this configuration the function of the moment would be graphed as follows:

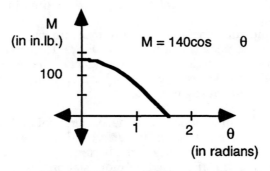

Figure 12.21 Plot of Moment vs. θ

Although the one-arm curl doesn't involve a full 360° cycle, the problem represents an example of the concept of *amplitude*. Sinusoidal functions involve the oscillation between maximum and minimum values. In this example, the magnitude of the moment fluctuates as a function of θ. The moment will be at its highest when $\theta = 0$. This is when the forearm is fully extended and the moment arm is the longest. When the forearm has rotated 90° ($\pi/2$ radians), the length of the moment arm is 0 and there is no moment. The moment will get no higher than 140 in.·lb. The value of 140 is therefore the amplitude of the function M = 140 cosθ.

The amplitude represents the maximum value of a sinusoidal function. Recall that the sine and cosine functions fluctuate between 1 and -1. When a coefficient is placed before the function, the amplitude of the sine wave is affected. The graphs of $y = 3\sin\theta$, $y = \sin\theta$, and $y = 1/3\sin\theta$ are shown in Figure 12.22.

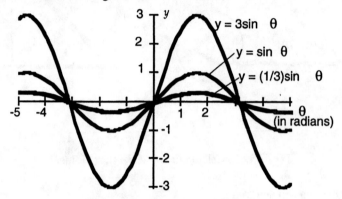

Figure 12.22 Amplitudes

Note that the amplitude of 3sinθ is three times that of sinθ. The amplitude of (1/3)sinθ is one third of sinθ. Also note that the three functions all cross the θ-axis at the same locations.

12.3.2 Frequency

The frequency of a cyclical function is one of its most important characteristics. In the example of an athlete jumping rope, we used the sine function to determine the height of the tip of the rope as a function of the angle of the rope. When working with rotating parts of machinery, designers are usually more interested in the position as a function of *time*. To get position as a function of time, we would need an equation of the form:

$$y = \sin t \qquad \text{where } t = \text{time, in seconds}$$

However the trigonometric functions can only measure *angles*. To get the sine as a function of time, we need a coefficient in front of the *t*. The coefficient which is usually used in cyclical (or *harmonic*) motion is ω. The product of the ωt term must have units of degrees or radians. This works because ω must have units of radians per unit time. This of course is the standard unit of angular velocity.

Quite often, the frequency of a motion is defined in terms of cycles per unit time. When the frequency is given in terms of cycles per second, it is usually denoted with the variable *f*. A common unit of frequency is the hertz (Hz). A hertz represents one cycle per second. The megahertz (MHz) used by radio stations to identify their frequencies represents a million cycles per second. The relationship between frequency, *f*, and angular velocity, ω, is given by the equation:

$$f = 2\pi\omega \qquad\qquad\qquad (12\text{–}14)$$

To be able to model cyclical motions with the trigonometric functions, one needs to be able to convert frequencies given in cycles per second to frequencies of radians per second. If an athlete jumps rope at 2 revolutions per second, the angular velocity is:

$$\left(\frac{2 \text{ rev.}}{\text{sec.}}\right)\left(\frac{2\pi \text{ radians}}{\text{rev.}}\right) = 4\pi \text{ radians}\Big/_{\text{sec.}}$$

The equation of $y = \sin 4\pi t$ \qquad where $\quad t$ = time, in seconds

y = the height of the rope, in meters

gives us the height of the rope relative to his hands at any time, t.

Figure 12.23 shows sine waves having different frequencies:

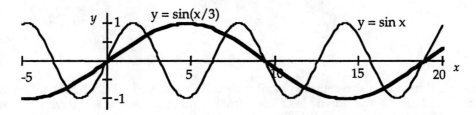

Figure 12.23 Different Frequencies

Note that the $\sin x$ function fits in three cycles by the time the $\sin(x/3)$ function has completed one cycle. The time it takes a trigonometric function to complete one cycle is known as its *period* and is usually denoted by the variable T. The period of a function is the inverse of its frequency. In mathematical terms:

$$f = \frac{1}{T} \qquad \text{and} \qquad T = \frac{1}{f} \qquad\qquad (12\text{-}14)$$

The period of the $\sin x$ function is 2π. Its frequency is therefore $1/2\pi$. The period of the $\sin(x/3)$ is three times the length of the period of the function $\sin x$ or 6π. If the period is three times that of $\sin x$, its frequency is one-third of $\sin x$. Note that the amplitudes of the functions remain the same when only the frequencies are affected.

The significance of different frequencies can be further explained by the example of hurdling. Hurdling could be viewed to involve two motions, running and the jumping of hurdles. The running strides are made at a particular frequency while the hurdling is done at a different frequency. A hurdler who 3-steps will make three strides with each leg between each hurdle. If a hurdler 3-steps with a stride at a frequency of x, the hurdling will have a frequency of $1/3x$.

12.3.3 Practical Examples

There are several examples of sinusoidal relationships in the physical world. When an object such as a satellite orbits a rotating planet, the projection of the satellite's path on the planet's surface is a sine wave.

Figure 12.24 Projection of Satellite Orbit

Sinusoidal motion is also common in vibration. If the spring in Fig. 12.25 is pulled down, it will oscillate up and down. As shown in the figure, its height will have a sinusoidal relationship with time until it is brought to rest. (When air friction slows a vibrating object, the vibration is said to be *dampened*. The affect of dampening causes a logarithmic reduction in amplitude.)

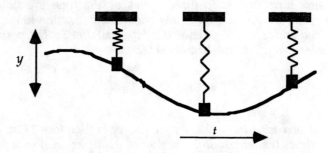

Figure 12.25 Spring Motion

Virtually any flexible object which that is moved out of its normal position can function like a spring. Figure 12.26 shows some objects that will vibrate if moved. As it is pulled in one direction, the object will attempt to return to its original position. As it springs back to its original position, it will gain some momentum and continue moving past its original position. It will reach a point where forces will again try to return it to its original position. It may take hundreds of cycles before the motion completely stops. This type of motion causes buildings to oscillate after an earthquake.

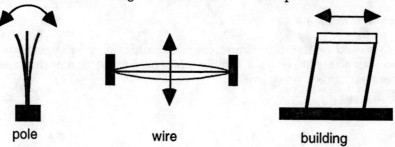

Figure 12.26 Other Types of Vibratory Motion

Oscillatory motion is also behind the creation of music from virtually any type of instrument. The strings on a violin, guitar, or piano will vibrate in the way described above. Drums, woodwind, and brass instruments all generate sound waves from materials vibrating back and forth.

Another type of vibration regularly occurs with moving machines and vehicles. Any type machinery that rotates or oscillates may vibrate in an undesirable way due to imperfections in its moving parts. For example, if the wheel on a car is out of alignment, it will create a vibration. The frequency of the vibration will equal the frequency of the wheel rotation. Similarly, if the blades on a ceiling fan are unbalanced, it will wobble when in use. A typical form of an equation describing the displacement of a vibrating object has the form:

$x = A\cos\omega t$ where x = the displacement of the object at time t (12 - 15)
                                       A = the maximum displacement
                                       ω = the frequency of the vibration, in radians per second
                                       t = time, in seconds

Another field in which sinusoidal oscillation is common is electronics. In an AC (alternating current) circuit, the components are arranged in such a way that as the current increases, physical forces are created which cause the current to reverse direction. After reversing direction, the current increases causing to again reverse direction. The process can happen thousands or millions of times per second. A common form of an equation to express the electrical current, i, of an AC circuit is:

$i = A\cos\omega t$ where i = the current at time t, in amperes   (12 - 16)
                                       A = the maximum current or amplitude, in amperes
                                       ω = the frequency of the current, in radians per second
                                       t = time, in seconds

As you might imagine, a great deal of trigonometry is required in the conversion between vibratory and electrical oscillations. A microphone is an instrument that converts vibratory oscillations (from an instrument or a person's voice) in to electrical oscillations. A speaker converts electrical oscillations into vibratory oscillations which produce sound.

The following problem illustrates an example of the significance of amplitude and frequency.

Example 12.9: Near the city of Acapulco, Mexico, there is a cliff which is a famous diving location. One of the diving points on the cliff is approximately 130 ft. above the water. Because the water is relatively shallow at this point, divers must time their dives so that they enter the water when the wave is the highest. Divers can see the point where they will enter the water. To dive safely, the diver needs to watch that point, wait for the crest of the wave to reach the point and then count a certain number of seconds to dive. If they jump at the right time, the wave will be near its crest again when they reach the water.

On a particular day, the tide places the depth of the water at the entrance point at 14 feet. The area, however, is experiencing sinusoidal waves which are 3 feet high from top to bottom. With the waves, the depth therefore fluctuates between 15.5 feet and 12.5 feet. The crests of the waves come every 8 seconds. To avoid risking injury from hitting the bottom, the diver should hit the water when it is at least 15 feet deep.

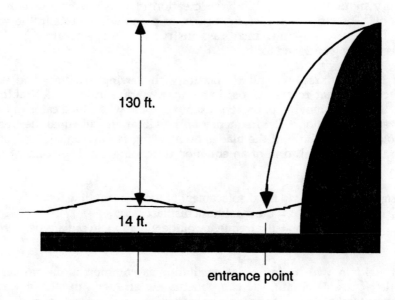

130 ft.

14 ft.

entrance point

Figure 12.27 Cliff Dive

Given the data, determine:

a) the time it takes the diver to reach the water at its average depth;
b) the depth of the water as a function of time;
c) the period of time or "window" during each wave when it is safe to enter the water (the time the water is at least 15 ft. deep);
d) the range of seconds after the crest passes the entrance point when the diver can safely dive.

Solution: a.) To determine the time it takes the diver to reach the water or t_d, we use the acceleration equation, Eq. 9-1. For $d = 130$ ft. and $a = g = 32.2$ ft./sec^2:

$$d = \frac{1}{2}at^2$$

$$t = \sqrt{\frac{2d}{a}}$$

$$= \sqrt{\frac{2(130 \text{ ft.})}{32.2 \text{ ft./sec.}^2}} = 2.84 \approx 3 \text{ sec.}$$

b.) If the height, h, of the wave were plotted as a function of time, the time periods could be graphed as shown in Figure 12.28.

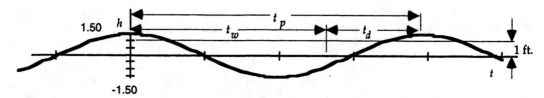

Figure 12.28 Time Periods between Waves

where:

t_p = time between crests, in seconds
t_w = waiting time, in seconds
t_d = diving time, in seconds

To get the depth of the water as a function of time, we need to find the height of the wave as a function of time. Since the problem states that the waves are sinusoidal, the height of the wave can be modeled with a sine or cosine function. The problem uses the occasions when the crests are at the entrance point as a reference point. Since the cosine function has its maximum (or crest) at $\theta = 0$, the cosine function fits the problem better. The equation for the depth of the water will take the form:

$y = 14 \pm$ the height of the wave or

$y = 14 + a\cos\omega t$ where y = the depth of the water, in ft.
    a = the amplitude of the wave, in ft.
    ω = the frequency of the waves, in radians per second

The amplitude of a wave form is the height from its zero point ($y = 0$) to its peak. If the problem defines the height of the wave as being from its top to bottom and its height is 3 ft., the amplitude of the wave is one-half of that or 1.5 feet. The height of the wave is therefore given by the equation $h = 1.5 \cos\omega t$.

If the crests come every 8 seconds, then each cycle has a period of 8 seconds. Therefore, there are 8 seconds per cycle. A cycle is defined as having 2π radians/cycle. To get ω in the required units (rad./sec.), we divide 2π rad./cycle by 8 sec./cycle.

$$\omega = \frac{2\pi\,\text{rad.}/\text{cycle}}{8\,\text{sec.}/\text{cycle}} \approx 0.8\ \text{rad.}/\text{sec.}$$

The equation for the depth of the water is therefore: $y = 14 + h$

$$= 14 + 1.5\cos0.8t$$

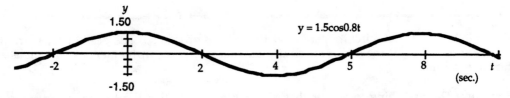

Figure 12.29 Height of Wave

c.) For the water to be 15 feet deep, the height of the wave needs to be 1 foot. To determine the values of t for $h = 1$ ft., we need to solve the equation for t.

$$h = 1.5\cos 0.8t = 1$$

$$\cos 0.8t = \frac{1}{1.5}$$

$$0.8t = \cos^{-1}\frac{1}{1.5}$$

$$t = \frac{\cos^{-1}\dfrac{1}{1.5}}{0.8}$$

$$= 1.05 \text{ sec.}$$

Since the cosine function is symmetrical about the y-axis, the wave will be at a height of exactly 1 ft. above the zero point at 1.05 seconds before and after the point where the wave is at its crest. There is therefore a "window" of approximately 2 seconds when it is safe to enter the water.

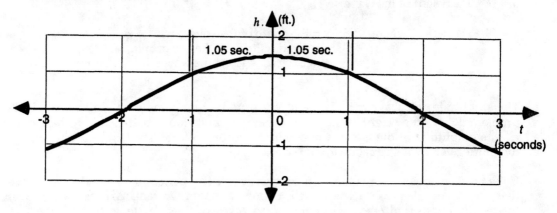

Figure 12.30 Safe Diving "Window"

d.) The range of seconds when it is safe to dive can be found by finding the time to dive to meet the center of the window (i.e., the crest of the wave) and then adding and subtracting 1 second. The number to count after seeing a crest at the entrance point is simply the period minus the diving time. The problem states that the times between waves, the period, is 8 seconds. Therefore, the waiting time to meet the crest equals 8 - 3 = 5 seconds. The range is therefore 5 ± 1 seconds or between 4 and 6 seconds after the crest of the wave passes the entrance point.

Example 12.10: An AC circuit operates at 10 amperes on 2 kHz. The peak of the current occurs at time = 0. If the current is sinusoidal, determine the instantaneous current as a function of time.

Solution: The first thing to note is that the problem states that the current is sinusoidal and that the peak current occurs at time = 0. If the peak occurs at time = 0, the function can be modeled with the cosine function. The equation will therefore have the form:

$$i = A\cos\omega t \quad \text{where}$$

i = the instantaneous current, in amperes
A = the amplitude of the current, in amperes
ω = the frequency of the current, in radians per second
t = time, in seconds

The problem states the current of the circuit, 10 amperes. Therefore the amplitude is 10. To determine, ω, we need to convert the frequency from hertz to radians per second.

$$\omega = \left(2\pi\,{}^{\text{rad.}}\!/_{\text{cycle}}\right)\left(2000\,{}^{\text{cycles}}\!/_{\text{second}}\right) \approx 12,566\,{}^{\text{rad.}}\!/_{\text{second}}$$

The instantaneous current of the circuit can therefore be given by the equation:

$$i = 10\cos 12,566t \qquad \text{with} \quad \begin{array}{l} i \text{ in amperes} \\ t \text{ in seconds} \end{array}$$

Exercise Set 12.3

Note - If you use a calculator to do these problems, make sure it's in the radians mode.

1. A basketball player bounces the ball with a constant rhythm having a period of 1.25 seconds. What is the frequency of the bouncing in bounces (cycles) per second?

2. An engine is operating at a frequency of 3,264 revolutions (cycles) per minute.
a.) What is the period length in minutes?
b.) What is the period length in seconds?
c.) What is the angular velocity in radians per second?

3. A part in the motor of a garbage disposal is out of alignment causing it to vibrate excessively. The resulting force on the sink is given by the equation:

$$F = 0.15\cos 60t \qquad \text{with } t \text{ in seconds and } F \text{ in lb.}$$

a.) What is the amplitude of the vibration, in lb.?
b.) What is the frequency of the vibration in Hz?
c.) What is the angular velocity of the vibration in radians per second?
d.) Plot the force of the garbage disposal as a function of time over three cycles.

4. The voltage through a particular component in an electrical circuit is given by the equation:

$$V = 400\cos 2000t \qquad \text{with} \quad \begin{array}{l} V \text{ in volts} \\ t \text{ in seconds} \end{array}$$

a.) What is the amplitude of the voltage?
b.) What is the frequency of the voltage in Hz?
c.) What is the frequency of the voltage in radians per second?

5. The oscillatory motion of a windshield wiper is converted from circular motion by the mechanism shown in Figure A. A spinning rotor causes a shaft to oscillate back and forth. A set of guides keeps the shaft straight. The shaft is attached to an arm which rotates a disc back and forth. A pin on the shaft slides in a groove on the arm as it oscillates. The disc rotates the wiper blade.

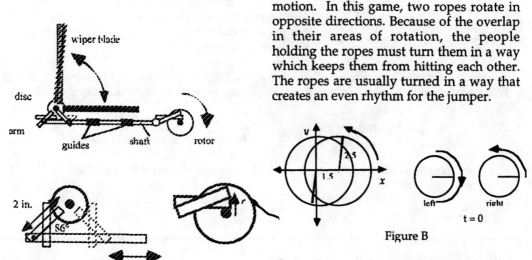

motion. In this game, two ropes rotate in opposite directions. Because of the overlap in their areas of rotation, the people holding the ropes must turn them in a way which keeps them from hitting each other. The ropes are usually turned in a way that creates an even rhythm for the jumper.

Figure B

Figure A

Specifications require that the wiper blade oscillate 86° back and forth at a rate of 1.2 cycles per second. The greatest distance between the center of the disc and the pin on the shaft is 2 in. Determine the required angular velocity ω and radius, *r*, of the rotor.

6. As mentioned at the beginning of the chapter, the double dutch style of jump rope involves a very complex oscillatory

The radius of the rotation at the point where the person jumps (in the middle) is 2.5 ft. The ropes are held 1.5 ft. apart from each other. Both ropes rotate at angular velocities of 8 radians/second but in opposite directions. The figure on the right shows the ropes separated at a reference point of time = 0. The origin is at the center of the rope on the left as shown in the figure. Determine the functions which give the *x* and *y* positions of the center of each rope as a function of time.

APPENDICES

APPENDIX A

COMMON LOGARITHMS TO THREE PLACES

	0.00	0.01	0.02	0.03	0.04	0.05	0.06	0.07	0.08	0.09
1.0	0.000	0.004	0.009	0.013	0.017	0.021	0.025	0.029	0.033	0.037
1.1	0.041	0.045	0.049	0.053	0.057	0.061	0.064	0.068	0.072	0.076
1.2	0.079	0.083	0.086	0.090	0.093	0.097	0.100	0.104	0.107	0.111
1.3	0.114	0.117	0.121	0.124	0.127	0.130	0.134	0.137	0.140	0.143
1.4	0.146	0.149	0.152	0.155	0.158	0.161	0.164	0.167	0.170	0.173
1.5	0.176	0.179	0.182	0.185	0.188	0.190	0.193	0.196	0.199	0.201
1.6	0.204	0.207	0.210	0.212	0.215	0.217	0.220	0.223	0.225	0.228
1.7	0.230	0.233	0.236	0.238	0.241	0.243	0.246	0.248	0.250	0.253
1.8	0.255	0.258	0.260	0.262	0.265	0.267	0.270	0.272	0.274	0.276
1.9	0.279	0.281	0.283	0.286	0.288	0.290	0.292	0.294	0.297	0.299
2.0	0.301	0.303	0.305	0.307	0.310	0.312	0.314	0.316	0.318	0.320
2.1	0.322	0.324	0.326	0.328	0.330	0.332	0.334	0.336	0.338	0.340
2.2	0.342	0.344	0.346	0.348	0.350	0.352	0.354	0.356	0.358	0.360
2.3	0.362	0.364	0.365	0.367	0.369	0.371	0.373	0.375	0.377	0.378
2.4	0.380	0.382	0.384	0.386	0.387	0.389	0.391	0.393	0.394	0.396
2.5	0.398	0.400	0.401	0.403	0.405	0.407	0.408	0.410	0.412	0.413
2.6	0.415	0.417	0.418	0.420	0.422	0.423	0.425	0.427	0.428	0.430
2.7	0.431	0.433	0.435	0.436	0.438	0.439	0.441	0.442	0.444	0.446
2.8	0.447	0.449	0.450	0.452	0.453	0.455	0.456	0.458	0.459	0.461
2.9	0.462	0.464	0.465	0.467	0.468	0.470	0.471	0.473	0.474	0.476
3.0	0.477	0.479	0.480	0.481	0.483	0.484	0.486	0.487	0.489	0.490
3.1	0.491	0.493	0.494	0.496	0.497	0.498	0.500	0.501	0.502	0.504
3.2	0.505	0.507	0.508	0.509	0.511	0.512	0.513	0.515	0.516	0.517
3.3	0.519	0.520	0.521	0.522	0.524	0.525	0.526	0.528	0.529	0.530
3.4	0.531	0.533	0.534	0.535	0.537	0.538	0.539	0.540	0.542	0.543
3.5	0.544	0.545	0.547	0.548	0.549	0.550	0.551	0.553	0.554	0.555
3.6	0.556	0.558	0.559	0.560	0.561	0.562	0.563	0.565	0.566	0.567
3.7	0.568	0.569	0.571	0.572	0.573	0.574	0.575	0.576	0.577	0.579
3.8	0.580	0.581	0.582	0.583	0.584	0.585	0.587	0.588	0.589	0.590
3.9	0.591	0.592	0.593	0.594	0.595	0.597	0.598	0.599	0.600	0.601
4.0	0.602	0.603	0.604	0.605	0.606	0.607	0.609	0.610	0.611	0.612
4.1	0.613	0.614	0.615	0.616	0.617	0.618	0.619	0.620	0.621	0.622
4.2	0.623	0.624	0.625	0.626	0.627	0.628	0.629	0.630	0.631	0.632
4.3	0.633	0.634	0.635	0.636	0.637	0.638	0.639	0.640	0.641	0.642
4.4	0.643	0.644	0.645	0.646	0.647	0.648	0.649	0.650	0.651	0.652
4.5	0.653	0.654	0.655	0.656	0.657	0.658	0.659	0.660	0.661	0.662
4.6	0.663	0.664	0.665	0.666	0.667	0.667	0.668	0.669	0.670	0.671
4.7	0.672	0.673	0.674	0.675	0.676	0.677	0.678	0.679	0.679	0.680
4.8	0.681	0.682	0.683	0.684	0.685	0.686	0.687	0.688	0.688	0.689
4.9	0.690	0.691	0.692	0.693	0.694	0.695	0.695	0.696	0.697	0.698
5.0	0.699	0.700	0.701	0.702	0.702	0.703	0.704	0.705	0.706	0.707
5.1	0.708	0.708	0.709	0.710	0.711	0.712	0.713	0.713	0.714	0.715
5.2	0.716	0.717	0.718	0.719	0.719	0.720	0.721	0.722	0.723	0.723
5.3	0.724	0.725	0.726	0.727	0.728	0.728	0.729	0.730	0.731	0.732
5.4	0.732	0.733	0.734	0.735	0.736	0.736	0.737	0.738	0.739	0.740

Appendix A Continued

COMMON LOGARITHMS TO THREE PLACES

	0.00	0.01	0.02	0.03	0.04	0.05	0.06	0.07	0.08	0.09
5.5	0.740	0.741	0.742	0.743	0.744	0.744	0.745	0.746	0.747	0.747
5.6	0.748	0.749	0.750	0.751	0.751	0.752	0.753	0.754	0.754	0.755
5.7	0.756	0.757	0.757	0.758	0.759	0.760	0.760	0.761	0.762	0.763
5.8	0.763	0.764	0.765	0.766	0.766	0.767	0.768	0.769	0.769	0.770
5.9	0.771	0.772	0.772	0.773	0.774	0.775	0.775	0.776	0.777	0.777
6.0	0.778	0.779	0.780	0.780	0.781	0.782	0.782	0.783	0.784	0.785
6.1	0.785	0.786	0.787	0.787	0.788	0.789	0.790	0.790	0.791	0.792
6.2	0.792	0.793	0.794	0.794	0.795	0.796	0.797	0.797	0.798	0.799
6.3	0.799	0.800	0.801	0.801	0.802	0.803	0.803	0.804	0.805	0.806
6.4	0.806	0.807	0.808	0.808	0.809	0.810	0.810	0.811	0.812	0.812
6.5	0.813	0.814	0.814	0.815	0.816	0.816	0.817	0.818	0.818	0.819
6.6	0.820	0.820	0.821	0.822	0.822	0.823	0.823	0.824	0.825	0.825
6.7	0.826	0.827	0.827	0.828	0.829	0.829	0.830	0.831	0.831	0.832
6.8	0.833	0.833	0.834	0.834	0.835	0.836	0.836	0.837	0.838	0.838
6.9	0.839	0.839	0.840	0.841	0.841	0.842	0.843	0.843	0.844	0.844
7.0	0.845	0.846	0.846	0.847	0.848	0.848	0.849	0.849	0.850	0.851
7.1	0.851	0.852	0.852	0.853	0.854	0.854	0.855	0.856	0.856	0.857
7.2	0.857	0.858	0.859	0.859	0.860	0.860	0.861	0.862	0.862	0.863
7.3	0.863	0.864	0.865	0.865	0.866	0.866	0.867	0.867	0.868	0.869
7.4	0.869	0.870	0.870	0.871	0.872	0.872	0.873	0.873	0.874	0.874
7.5	0.875	0.876	0.876	0.877	0.877	0.878	0.879	0.879	0.880	0.880
7.6	0.881	0.881	0.882	0.883	0.883	0.884	0.884	0.885	0.885	0.886
7.7	0.886	0.887	0.888	0.888	0.889	0.889	0.890	0.890	0.891	0.892
7.8	0.892	0.893	0.893	0.894	0.894	0.895	0.895	0.896	0.897	0.897
7.9	0.898	0.898	0.899	0.899	0.900	0.900	0.901	0.901	0.902	0.903
8.0	0.903	0.904	0.904	0.905	0.905	0.906	0.906	0.907	0.907	0.908
8.1	0.908	0.909	0.910	0.910	0.911	0.911	0.912	0.912	0.913	0.913
8.2	0.914	0.914	0.915	0.915	0.916	0.916	0.917	0.918	0.918	0.919
8.3	0.919	0.920	0.920	0.921	0.921	0.922	0.922	0.923	0.923	0.924
8.4	0.924	0.925	0.925	0.926	0.926	0.927	0.927	0.928	0.928	0.929
8.5	0.929	0.930	0.930	0.931	0.931	0.932	0.932	0.933	0.933	0.934
8.6	0.934	0.935	0.936	0.936	0.937	0.937	0.938	0.938	0.939	0.939
8.7	0.940	0.940	0.941	0.941	0.942	0.942	0.943	0.943	0.943	0.944
8.8	0.944	0.945	0.945	0.946	0.946	0.947	0.947	0.948	0.948	0.949
8.9	0.949	0.950	0.950	0.951	0.951	0.952	0.952	0.953	0.953	0.954
9.0	0.954	0.955	0.955	0.956	0.956	0.957	0.957	0.958	0.958	0.959
9.1	0.959	0.960	0.960	0.960	0.961	0.961	0.962	0.962	0.963	0.963
9.2	0.964	0.964	0.965	0.965	0.966	0.966	0.967	0.967	0.968	0.968
9.3	0.968	0.969	0.969	0.970	0.970	0.971	0.971	0.972	0.972	0.973
9.4	0.973	0.974	0.974	0.975	0.975	0.975	0.976	0.976	0.977	0.977
9.5	0.978	0.978	0.979	0.979	0.980	0.980	0.980	0.981	0.981	0.982
9.6	0.982	0.983	0.983	0.984	0.984	0.985	0.985	0.985	0.986	0.986
9.7	0.987	0.987	0.988	0.988	0.989	0.989	0.989	0.990	0.990	0.991
9.8	0.991	0.992	0.992	0.993	0.993	0.993	0.994	0.994	0.995	0.995
9.9	0.996	0.996	0.997	0.997	0.997	0.998	0.998	0.999	0.999	1.000

APPENDIX B

TRIGONOMETRIC FUNCTIONS

Degrees	Radians	Sine	Cosine	Tangent	Cotangent	Secant	Cosecant
0.0	0.0000	0.0000	1.0000	0.0000	undefined	1.0000	undefined
1.0	0.0175	0.0175	0.9998	0.0175	57.2900	1.0002	57.2987
2.0	0.0349	0.0349	0.9994	0.0349	28.6363	1.0006	28.6537
3.0	0.0524	0.0523	0.9986	0.0524	19.0811	1.0014	19.1073
4.0	0.0698	0.0698	0.9976	0.0699	14.3007	1.0024	14.3356
5.0	0.0873	0.0872	0.9962	0.0875	11.4301	1.0038	11.4737
6.0	0.1047	0.1045	0.9945	0.1051	9.5144	1.0055	9.5668
7.0	0.1222	0.1219	0.9925	0.1228	8.1443	1.0075	8.2055
8.0	0.1396	0.1392	0.9903	0.1405	7.1154	1.0098	7.1853
9.0	0.1571	0.1564	0.9877	0.1584	6.3138	1.0125	6.3925
10.0	0.1745	0.1736	0.9848	0.1763	5.6713	1.0154	5.7588
11.0	0.1920	0.1908	0.9816	0.1944	5.1446	1.0187	5.2408
12.0	0.2094	0.2079	0.9781	0.2126	4.7046	1.0223	4.8097
13.0	0.2269	0.2250	0.9744	0.2309	4.3315	1.0263	4.4454
14.0	0.2443	0.2419	0.9703	0.2493	4.0108	1.0306	4.1336
15.0	0.2618	0.2588	0.9659	0.2679	3.7321	1.0353	3.8637
16.0	0.2793	0.2756	0.9613	0.2867	3.4874	1.0403	3.6280
17.0	0.2967	0.2924	0.9563	0.3057	3.2709	1.0457	3.4203
18.0	$\pi/10$	0.3090	0.9511	0.3249	3.0777	1.0515	3.2361
19.0	0.3316	0.3256	0.9455	0.3443	2.9042	1.0576	3.0716
20.0	$\pi/9$	0.3420	0.9397	0.3640	2.7475	1.0642	2.9238
21.0	0.3665	0.3584	0.9336	0.3839	2.6051	1.0711	2.7904
22.0	0.3840	0.3746	0.9272	0.4040	2.4751	1.0785	2.6695
23.0	0.4014	0.3907	0.9205	0.4245	2.3559	1.0864	2.5593
24.0	0.4189	0.4067	0.9135	0.4452	2.2460	1.0946	2.4586
25.0	0.4363	0.4226	0.9063	0.4663	2.1445	1.1034	2.3662
26.0	0.4538	0.4384	0.8988	0.4877	2.0503	1.1126	2.2812
27.0	0.4712	0.4540	0.8910	0.5095	1.9626	1.1223	2.2027
28.0	0.4887	0.4695	0.8829	0.5317	1.8807	1.1326	2.1301
29.0	0.5061	0.4848	0.8746	0.5543	1.8040	1.1434	2.0627
30.0	$\pi/6$	0.5000	0.8660	0.5774	1.7321	1.1547	2.0000
31.0	0.5411	0.5150	0.8572	0.6009	1.6643	1.1666	1.9416
32.0	0.5585	0.5299	0.8480	0.6249	1.6003	1.1792	1.8871
33.0	0.5760	0.5446	0.8387	0.6494	1.5399	1.1924	1.8361
34.0	0.5934	0.5592	0.8290	0.6745	1.4826	1.2062	1.7883
35.0	0.6109	0.5736	0.8192	0.7002	1.4281	1.2208	1.7434
36.0	$\pi/5$	0.5878	0.8090	0.7265	1.3764	1.2361	1.7013
37.0	0.6458	0.6018	0.7986	0.7536	1.3270	1.2521	1.6616
38.0	0.6632	0.6157	0.7880	0.7813	1.2799	1.2690	1.6243
39.0	0.6807	0.6293	0.7771	0.8098	1.2349	1.2868	1.5890
40.0	$2\pi/9$	0.6428	0.7660	0.8391	1.1918	1.3054	1.5557
41.0	0.7156	0.6561	0.7547	0.8693	1.1504	1.3250	1.5243
42.0	0.7330	0.6691	0.7431	0.9004	1.1106	1.3456	1.4945
43.0	0.7505	0.6820	0.7314	0.9325	1.0724	1.3673	1.4663
44.0	0.7679	0.6947	0.7193	0.9657	1.0355	1.3902	1.4396
45.0	$\pi/4$	0.7071	0.7071	1.0000	1.0000	1.4142	1.4142

Appendix B Continued

TRIGONOMETRIC FUNCTIONS

Degrees	Radians	Sine	Cosine	Tangent	Cotangent	Secant	Cosecant
46.0	0.8029	0.7193	0.6947	1.0355	0.9657	1.4396	1.3902
47.0	0.8203	0.7314	0.6820	1.0724	0.9325	1.4663	1.3673
48.0	0.8378	0.7431	0.6691	1.1106	0.9004	1.4945	1.3456
49.0	0.8552	0.7547	0.6561	1.1504	0.8693	1.5243	1.3250
50.0	0.8727	0.7660	0.6428	1.1918	0.8391	1.5557	1.3054
51.0	0.8901	0.7771	0.6293	1.2349	0.8098	1.5890	1.2868
52.0	0.9076	0.7880	0.6157	1.2799	0.7813	1.6243	1.2690
53.0	0.9250	0.7986	0.6018	1.3270	0.7536	1.6616	1.2521
54.0	$3\pi/10$	0.8090	0.5878	1.3764	0.7265	1.7013	1.2361
55.0	0.9599	0.8192	0.5736	1.4281	0.7002	1.7434	1.2208
56.0	0.9774	0.8290	0.5592	1.4826	0.6745	1.7883	1.2062
57.0	0.9948	0.8387	0.5446	1.5399	0.6494	1.8361	1.1924
58.0	1.0123	0.8480	0.5299	1.6003	0.6249	1.8871	1.1792
59.0	1.0297	0.8572	0.5150	1.6643	0.6009	1.9416	1.1666
60.0	$\pi/3$	0.8660	0.5000	1.7321	0.5774	2.0000	1.1547
61.0	1.0647	0.8746	0.4848	1.8040	0.5543	2.0627	1.1434
62.0	1.0821	0.8829	0.4695	1.8807	0.5317	2.1301	1.1326
63.0	1.0996	0.8910	0.4540	1.9626	0.5095	2.2027	1.1223
64.0	1.1170	0.8988	0.4384	2.0503	0.4877	2.2812	1.1126
65.0	1.1345	0.9063	0.4226	2.1445	0.4663	2.3662	1.1034
66.0	1.1519	0.9135	0.4067	2.2460	0.4452	2.4586	1.0946
67.0	1.1694	0.9205	0.3907	2.3559	0.4245	2.5593	1.0864
68.0	1.1868	0.9272	0.3746	2.4751	0.4040	2.6695	1.0785
69.0	1.2043	0.9336	0.3584	2.6051	0.3839	2.7904	1.0711
70.0	1.2217	0.9397	0.3420	2.7475	0.3640	2.9238	1.0642
71.0	1.2392	0.9455	0.3256	2.9042	0.3443	3.0716	1.0576
72.0	$2\pi/5$	0.9511	0.3090	3.0777	0.3249	3.2361	1.0515
73.0	1.2741	0.9563	0.2924	3.2709	0.3057	3.4203	1.0457
74.0	1.2915	0.9613	0.2756	3.4874	0.2867	3.6280	1.0403
75.0	1.3090	0.9659	0.2588	3.7321	0.2679	3.8637	1.0353
76.0	1.3265	0.9703	0.2419	4.0108	0.2493	4.1336	1.0306
77.0	1.3439	0.9744	0.2250	4.3315	0.2309	4.4454	1.0263
78.0	1.3614	0.9781	0.2079	4.7046	0.2126	4.8097	1.0223
79.0	1.3788	0.9816	0.1908	5.1446	0.1944	5.2408	1.0187
80.0	$4\pi/9$	0.9848	0.1736	5.6713	0.1763	5.7588	1.0154
81.0	1.4137	0.9877	0.1564	6.3138	0.1584	6.3925	1.0125
82.0	1.4312	0.9903	0.1392	7.1154	0.1405	7.1853	1.0098
83.0	1.4486	0.9925	0.1219	8.1443	0.1228	8.2055	1.0075
84.0	1.4661	0.9945	0.1045	9.5144	0.1051	9.5668	1.0055
85.0	1.4835	0.9962	0.0872	11.4301	0.0875	11.4737	1.0038
86.0	1.5010	0.9976	0.0698	14.3007	0.0699	14.3356	1.0024
87.0	1.5184	0.9986	0.0523	19.0811	0.0524	19.1073	1.0014
88.0	1.5359	0.9994	0.0349	28.6363	0.0349	28.6537	1.0006
89.0	1.5533	0.9998	0.0175	57.2900	0.0175	57.2987	1.0002
90.0	$\pi/2$	1.0000	0.0000	undefined	0.0000	undefined	1.0000

Appendix B Continued

TRIGONOMETRIC FUNCTIONS

Degrees	Radians	Sine	Cosine	Tangent	Cotangent	Secant	Cosecant
91.0	1.5882	0.9998	-0.0175	-57.29	-0.0175	-57.299	1.0002
92.0	1.6057	0.9994	-0.0349	-28.636	-0.0349	-28.654	1.0006
93.0	1.6232	0.9986	-0.0523	-19.081	-0.0524	-19.107	1.0014
94.0	1.6406	0.9976	-0.0698	-14.301	-0.0699	-14.336	1.0024
95.0	1.6581	0.9962	-0.0872	-11.43	-0.0875	-11.474	1.0038
96.0	1.6755	0.9945	-0.1045	-9.5144	-0.1051	-9.5668	1.0055
97.0	1.6930	0.9925	-0.1219	-8.1443	-0.1228	-8.2055	1.0075
98.0	1.7104	0.9903	-0.1392	-7.1154	-0.1405	-7.1853	1.0098
99.0	1.7279	0.9877	-0.1564	-6.3138	-0.1584	-6.3925	1.0125
100.0	$5\pi/9$	0.9848	-0.1736	-5.6713	-0.1763	-5.7588	1.0154
101.0	1.7628	0.9816	-0.1908	-5.1446	-0.1944	-5.2408	1.0187
102.0	1.7802	0.9781	-0.2079	-4.7046	-0.2126	-4.8097	1.0223
103.0	1.7977	0.9744	-0.2250	-4.3315	-0.2309	-4.4454	1.0263
104.0	1.8151	0.9703	-0.2419	-4.0108	-0.2493	-4.1336	1.0306
105.0	1.8326	0.9659	-0.2588	-3.7321	-0.2679	-3.8637	1.0353
106.0	1.8500	0.9613	-0.2756	-3.4874	-0.2867	-3.6280	1.0403
107.0	1.8675	0.9563	-0.2924	-3.2709	-0.3057	-3.4203	1.0457
108.0	$3\pi/5$	0.9511	-0.3090	-3.0777	-0.3249	-3.2361	1.0515
109.0	1.9024	0.9455	-0.3256	-2.9042	-0.3443	-3.0716	1.0576
110.0	1.9199	0.9397	-0.3420	-2.7475	-0.3640	-2.9238	1.0642
111.0	1.9373	0.9336	-0.3584	-2.6051	-0.3839	-2.7904	1.0711
112.0	1.9548	0.9272	-0.3746	-2.4751	-0.4040	-2.6695	1.0785
113.0	1.9722	0.9205	-0.3907	-2.3559	-0.4245	-2.5593	1.0864
114.0	1.9897	0.9135	-0.4067	-2.2460	-0.4452	-2.4586	1.0946
115.0	2.0071	0.9063	-0.4226	-2.1445	-0.4663	-2.3662	1.1034
116.0	2.0246	0.8988	-0.4384	-2.0503	-0.4877	-2.2812	1.1126
117.0	2.0420	0.8910	-0.4540	-1.9626	-0.5095	-2.2027	1.1223
118.0	2.0595	0.8829	-0.4695	-1.8807	-0.5317	-2.1301	1.1326
119.0	2.0769	0.8746	-0.4848	-1.8040	-0.5543	-2.0627	1.1434
120.0	$2\pi/3$	0.8660	-0.5000	-1.7321	-0.5774	-2.0000	1.1547
121.0	2.1118	0.8572	-0.5150	-1.6643	-0.6009	-1.9416	1.1666
122.0	2.1293	0.8480	-0.5299	-1.6003	-0.6249	-1.8871	1.1792
123.0	2.1468	0.8387	-0.5446	-1.5399	-0.6494	-1.8361	1.1924
124.0	2.1642	0.8290	-0.5592	-1.4826	-0.6745	-1.7883	1.2062
125.0	2.1817	0.8192	-0.5736	-1.4281	-0.7002	-1.7434	1.2208
126.0	$7\pi/10$	0.8090	-0.5878	-1.3764	-0.7265	-1.7013	1.2361
127.0	2.2166	0.7986	-0.6018	-1.3270	-0.7536	-1.6616	1.2521
128.0	2.2340	0.7880	-0.6157	-1.2799	-0.7813	-1.6243	1.2690
129.0	2.2515	0.7771	-0.6293	-1.2349	-0.8098	-1.5890	1.2868
130.0	2.2689	0.7660	-0.6428	-1.1918	-0.8391	-1.5557	1.3054
131.0	2.2864	0.7547	-0.6561	-1.1504	-0.8693	-1.5243	1.3250
132.0	2.3038	0.7431	-0.6691	-1.1106	-0.9004	-1.4945	1.3456
133.0	2.3213	0.7314	-0.6820	-1.0724	-0.9325	-1.4663	1.3673
134.0	2.3387	0.7193	-0.6947	-1.0355	-0.9657	-1.4396	1.3902
135.0	$3\pi/4$	0.7071	-0.7071	-1.0000	-1.0000	-1.4142	1.4142

Appendix B Continued

TRIGONOMETRIC FUNCTIONS

Degrees	Radians	Sine	Cosine	Tangent	Cotangent	Secant	Cosecant
136.0	2.3736	0.6947	-0.7193	-0.9657	-1.0355	-1.3902	1.4396
137.0	2.3911	0.6820	-0.7314	-0.9325	-1.0724	-1.3673	1.4663
138.0	2.4086	0.6691	-0.7431	-0.9004	0.0000	-1.3456	1.4945
139.0	2.4260	0.6561	-0.7547	-0.8693	-1.1504	-1.325	1.5243
140.0	$7\pi/9$	0.6428	-0.7660	-0.8391	-1.1918	-1.3054	1.5557
141.0	2.4609	0.6293	-0.7771	-0.8098	-1.2349	-1.2868	1.5890
142.0	2.4784	0.6157	-0.7880	-0.7813	-1.2799	-1.269	1.6243
143.0	2.4958	0.6018	-0.7986	-0.7536	-1.3270	-1.2521	1.6616
144.0	$4\pi/5$	0.5878	-0.8090	-0.7265	-1.3764	-1.2361	1.7013
145.0	2.5307	0.5736	-0.8192	-0.7002	-1.4281	-1.2208	1.7434
146.0	2.5482	0.5592	-0.8290	-0.6745	-1.4826	-1.2062	1.7883
147.0	2.5656	0.5446	-0.8387	-0.6494	-1.5399	-1.1924	1.8361
148.0	2.5831	0.5299	-0.8480	-0.6249	-1.6003	-1.1792	1.8871
149.0	2.6005	0.5150	-0.8572	-0.6009	-1.6643	-1.1666	1.9416
150.0	$5\pi/6$	0.5000	-0.8660	-0.5774	-1.7321	-1.1547	2.0000
151.0	2.6354	0.4848	-0.8746	-0.5543	-1.8040	-1.1434	2.0627
152.0	2.6529	0.4695	-0.8829	-0.5317	-1.8807	-1.1326	2.1301
153.0	2.6704	0.4540	-0.8910	-0.5095	-1.9626	-1.1223	2.2027
154.0	2.6878	0.4384	-0.8988	-0.4877	-2.0503	-1.1126	2.2812
155.0	2.7053	0.4226	-0.9063	-0.4663	-2.1445	-1.1034	2.3662
156.0	2.7227	0.4067	-0.9135	-0.4452	-2.2460	-1.0946	2.4586
157.0	2.7402	0.3907	-0.9205	-0.4245	-2.3559	-1.0864	2.5593
158.0	2.7576	0.3746	-0.9272	-0.4040	-2.4751	-1.0785	2.6695
159.0	2.7751	0.3584	-0.9336	-0.3839	-2.6051	-1.0711	2.7904
160.0	$8\pi/9$	0.3420	-0.9397	-0.3640	-2.7475	-1.0642	2.9238
161.0	2.8100	0.3256	-0.9455	-0.3443	-2.9042	-1.0576	3.0716
162.0	$9\pi/10$	0.3090	-0.9511	-0.3249	-3.0777	-1.0515	3.2361
163.0	2.8449	0.2924	-0.9563	-0.3057	-3.2709	-1.0457	3.4203
164.0	2.8623	0.2756	-0.9613	-0.2867	-3.4874	-1.0403	3.6280
165.0	2.8798	0.2588	-0.9659	-0.2679	-3.7321	-1.0353	3.8637
166.0	2.8972	0.2419	-0.9703	-0.2493	-4.0108	-1.0306	4.1336
167.0	2.9147	0.2250	-0.9744	-0.2309	-4.3315	-1.0263	4.4454
168.0	2.9322	0.2079	-0.9781	-0.2126	-4.7046	-1.0223	4.8097
169.0	2.9496	0.1908	-0.9816	-0.1944	-5.1446	-1.0187	5.2408
170.0	2.9671	0.1736	-0.9848	-0.1763	-5.6713	-1.0154	5.7588
171.0	2.9845	0.1564	-0.9877	-0.1584	-6.3138	-1.0125	6.3925
172.0	3.0020	0.1392	-0.9903	-0.1405	-7.1154	-1.0098	7.1853
173.0	3.0194	0.1219	-0.9925	-0.1228	-8.1443	-1.0075	8.2055
174.0	3.0369	0.1045	-0.9945	-0.1051	-9.5144	-1.0055	9.5668
175.0	3.0543	0.0872	-0.9962	-0.0875	-11.43	-1.0038	11.4737
176.0	3.0718	0.0698	-0.9976	-0.0699	-14.301	-1.0024	14.3356
177.0	3.0892	0.0523	-0.9986	-0.0524	-19.081	-1.0014	19.1073
178.0	3.1067	0.0349	-0.9994	-0.0349	-28.636	-1.0006	28.6537
179.0	3.1241	0.0175	-0.9998	-0.0175	-57.29	-1.0002	57.2987
180.0	π	0.0000	-1.0000	0.0000	undefined	-1.0000	undefined

Appendix B Continued

TRIGONOMETRIC FUNCTIONS

Degrees	Radians	Sine	Cosine	Tangent	Cotangent	Secant	Cosecant
181.0	3.1590	-0.0175	-0.9998	0.0175	57.2900	-1.0002	-57.299
182.0	3.1765	-0.0349	-0.9994	0.0349	28.6363	-1.0006	-28.654
183.0	3.1940	-0.0523	-0.9986	0.0524	19.0811	-1.0014	-19.107
184.0	3.2114	-0.0698	-0.9976	0.0699	14.3007	-1.0024	-14.336
185.0	3.2289	-0.0872	-0.9962	0.0875	11.4301	-1.0038	-11.474
186.0	3.2463	-0.1045	-0.9945	0.1051	0.0000	-1.0055	-9.5668
187.0	3.2638	-0.1219	-0.9925	0.1228	8.1443	-1.0075	-8.2055
188.0	3.2812	-0.1392	-0.9903	0.1405	7.1154	-1.0098	-7.1853
189.0	3.2987	-0.1564	-0.9877	0.1584	6.3138	-1.0125	-6.3925
190.0	3.3161	-0.1736	-0.9848	0.1763	5.6713	-1.0154	-5.7588
191.0	3.3336	-0.1908	-0.9816	0.1944	5.1446	-1.0187	-5.2408
192.0	3.3510	-0.2079	-0.9781	0.2126	4.7046	-1.0223	-4.8097
193.0	3.3685	-0.2250	-0.9744	0.2309	4.3315	-1.0263	-4.4454
194.0	3.3859	-0.2419	-0.9703	0.2493	4.0108	-1.0306	-4.1336
195.0	3.4034	-0.2588	-0.9659	0.2679	3.7321	-1.0353	-3.8637
196.0	3.4208	-0.2756	-0.9613	0.2867	3.4874	-1.0403	-3.6280
197.0	3.4383	-0.2924	-0.9563	0.3057	3.2709	-1.0457	-3.4203
198.0	$11\pi/10$	-0.3090	-0.9511	0.3249	3.0777	-1.0515	-3.2361
199.0	3.4732	-0.3256	-0.9455	0.3443	2.9042	-1.0576	-3.0716
200.0	$10\pi/9$	-0.3420	-0.9397	0.3640	2.7475	-1.0642	-2.9238
201.0	3.5081	-0.3584	-0.9336	0.3839	2.6051	-1.0711	-2.7904
202.0	3.5256	-0.3746	-0.9272	0.4040	2.4751	-1.0785	-2.6695
203.0	3.5430	-0.3907	-0.9205	0.4245	2.3559	-1.0864	-2.5593
204.0	3.5605	-0.4067	-0.9135	0.4452	2.2460	-1.0946	-2.4586
205.0	3.5779	-0.4226	-0.9063	0.4663	2.1445	-1.1034	-2.3662
206.0	3.5954	-0.4384	-0.8988	0.4877	2.0503	-1.1126	-2.2812
207.0	3.6128	-0.4540	-0.8910	0.5095	1.9626	-1.1223	-2.2027
208.0	3.6303	-0.4695	-0.8829	0.5317	1.8807	-1.1326	-2.1301
209.0	3.6477	-0.4848	-0.8746	0.5543	1.8040	-1.1434	-2.0627
210.0	$7\pi/6$	-0.5000	-0.8660	0.5774	1.7321	-1.1547	-2.0000
211.0	3.6826	-0.5150	-0.8572	0.6009	1.6643	-1.1666	-1.9416
212.0	3.7001	-0.5299	-0.8480	0.6249	1.6003	-1.1792	-1.8871
213.0	3.7176	-0.5446	-0.8387	0.6494	1.5399	-1.1924	-1.8361
214.0	3.7350	-0.5592	-0.8290	0.6745	1.4826	-1.2062	-1.7883
215.0	3.7525	-0.5736	-0.8192	0.7002	1.4281	-1.2208	-1.7434
216.0	$6\pi/5$	-0.5878	-0.8090	0.7265	1.3764	-1.2361	-1.7013
217.0	3.7874	-0.6018	-0.7986	0.7536	1.3270	-1.2521	-1.6616
218.0	3.8048	-0.6157	-0.7880	0.7813	1.2799	-1.2690	-1.6243
219.0	3.8223	-0.6293	-0.7771	0.8098	1.2349	-1.2868	-1.5890
220.0	$11\pi/9$	-0.6428	-0.7660	0.8391	1.1918	-1.3054	-1.5557
221.0	3.8572	-0.6561	-0.7547	0.8693	1.1504	-1.3250	-1.5243
222.0	3.8746	-0.6691	-0.7431	0.9004	1.1106	-1.3456	-1.4945
223.0	3.8921	-0.6820	-0.7314	0.9325	1.0724	-1.3673	-1.4663
224.0	3.9095	-0.6947	-0.7193	0.9657	1.0355	-1.3902	-1.4396
225.0	$5\pi/4$	-0.7071	-0.7071	1.0000	1.0000	-1.4142	-1.4142

Appendix B Continued

TRIGONOMETRIC FUNCTIONS

Degrees	Radians	Sine	Cosine	Tangent	Cotangent	Secant	Cosecant
226.0	3.9444	-0.7193	-0.6947	1.0355	0.9657	-1.4396	-1.3902
227.0	3.9619	-0.7314	-0.6820	1.0724	0.9325	-1.4663	-1.3673
228.0	3.9794	-0.7431	-0.6691	1.1106	0.9004	-1.4945	-1.3456
229.0	3.9968	-0.7547	-0.6561	1.1504	0.8693	-1.5243	-1.3250
230.0	4.0143	-0.7660	-0.6428	1.1918	0.8391	-1.5557	-1.3054
231.0	4.0317	-0.7771	-0.6293	1.2349	0.8098	-1.5890	-1.2868
232.0	4.0492	-0.7880	-0.6157	1.2799	0.7813	-1.6243	-1.2690
233.0	4.0666	-0.7986	-0.6018	1.3270	0.7536	-1.6616	-1.2521
234.0	$13\pi/10$	-0.8090	-0.5878	1.3764	0.7265	-1.7013	-1.2361
235.0	4.1015	-0.8192	-0.5736	1.4281	0.7002	-1.7434	-1.2208
236.0	4.1190	-0.8290	-0.5592	1.4826	0.6745	-1.7883	-1.2062
237.0	4.1364	-0.8387	-0.5446	1.5399	0.6494	-1.8361	-1.1924
238.0	4.1539	-0.8480	-0.5299	1.6003	0.6249	-1.8871	-1.1792
239.0	4.1713	-0.8572	-0.5150	1.66428	0.6009	-1.9416	-1.1666
240.0	$4\pi/3$	-0.8660	-0.5000	1.7321	0.5774	-2.0000	-1.1547
241.0	4.2062	-0.8746	-0.4848	1.8040	0.5543	-2.0627	-1.1434
242.0	4.2237	-0.8829	-0.4695	1.8807	0.5317	-2.1301	-1.1326
243.0	4.2412	-0.8910	-0.4540	1.9626	0.5095	-2.2027	-1.1223
244.0	4.2586	-0.8988	-0.4384	2.0503	0.4877	-2.2812	-1.1126
245.0	4.2761	-0.9063	-0.4226	2.1445	0.4663	-2.3662	-1.1034
246.0	4.2935	-0.9135	-0.4067	2.2460	0.4452	-2.4586	-1.0946
247.0	4.3110	-0.9205	-0.3907	2.3559	0.4245	-2.5593	-1.0864
248.0	4.3284	-0.9272	-0.3746	2.4751	0.4040	-2.6695	-1.0785
249.0	4.3459	-0.9336	-0.3584	2.6051	0.3839	-2.7904	-1.0711
250.0	4.3633	-0.9397	-0.3420	2.7475	0.3640	-2.9238	-1.0642
251.0	4.3808	-0.9455	-0.3256	2.9042	0.3443	-3.0716	-1.0576
252.0	$7\pi/5$	-0.9511	-0.3090	3.0777	0.3249	-3.2361	-1.0515
253.0	4.4157	-0.9563	-0.2924	3.2709	0.3057	-3.4203	-1.0457
254.0	4.4331	-0.9613	-0.2756	3.4874	0.2867	-3.6280	-1.0403
255.0	4.4506	-0.9659	-0.2588	3.7321	0.2679	-3.8637	-1.0353
256.0	4.4680	-0.9703	-0.2419	4.0108	0.2493	-4.1336	-1.0306
257.0	4.4855	-0.9744	-0.2250	4.3315	0.2309	-4.4454	-1.0263
258.0	4.5029	-0.9781	-0.2079	4.7046	0.2126	-4.8097	-1.0223
259.0	4.5204	-0.9816	-0.1908	5.1446	0.1944	-5.2408	-1.0187
260.0	$13\pi/9$	-0.9848	-0.1736	5.6713	0.1763	-5.7588	-1.0154
261.0	4.5553	-0.9877	-0.1564	6.3138	0.1584	-6.3925	-1.0125
262.0	4.5728	-0.9903	-0.1392	7.1154	0.1405	-7.1853	-1.0098
263.0	4.5902	-0.9925	-0.1219	8.1443	0.1228	-8.2055	-1.0075
264.0	4.6077	-0.9945	-0.1045	9.5144	0.1051	-9.5668	-1.0055
265.0	4.6251	-0.9962	-0.0872	11.4301	0.0875	-11.474	-1.0038
266.0	4.6426	-0.9976	-0.0698	14.3007	0.0699	-14.336	-1.0024
267.0	4.6600	-0.9986	-0.0523	19.0811	0.0524	-19.107	-1.0014
268.0	4.6775	-0.9994	-0.0349	28.6363	0.0349	-28.654	-1.0006
269.0	4.6949	-0.9998	-0.0175	57.2900	0.0175	-57.299	-1.0002
270.0	$3\pi/2$	-1.0000	0.0000	undefined	0.0000	undefined	-1.0000

Appendix B Continued

TRIGONOMETRIC FUNCTIONS

Degrees	Radians	Sine	Cosine	Tangent	Cotangent	Secant	Cosecant
271.0	4.7298	-0.9998	0.0175	-57.29	-0.0175	57.2987	-1.0002
272.0	4.7473	-0.9994	0.0349	-28.636	-0.0349	28.6537	-1.0006
273.0	4.7647	-0.9986	0.0523	-19.081	-0.0524	19.1073	-1.0014
274.0	4.7822	-0.9976	0.0698	-14.301	-0.0699	14.3356	-1.0024
275.0	4.7997	-0.9962	0.0872	-11.43	-0.0875	11.4737	-1.0038
276.0	4.8171	-0.9945	0.1045	-9.5144	-0.1051	9.5668	-1.0055
277.0	4.8346	-0.9925	0.1219	-8.1443	-0.1228	8.2055	-1.0075
278.0	4.8520	-0.9903	0.1392	-7.1154	-0.1405	7.1853	-1.0098
279.0	4.8695	-0.9877	0.1564	-6.3138	-0.1584	6.3925	-1.0125
280.0	$14\pi/9$	-0.9848	0.1736	-5.6713	-0.1763	5.7588	-1.0154
281.0	4.9044	-0.9816	0.1908	-5.1446	-0.1944	5.2408	-1.0187
282.0	4.9218	-0.9781	0.2079	-4.7046	-0.2126	4.8097	-1.0223
283.0	4.9393	-0.9744	0.2250	-4.3315	-0.2309	4.4454	-1.0263
284.0	4.9567	-0.9703	0.2419	-4.0108	-0.2493	4.1336	-1.0306
285.0	4.9742	-0.9659	0.2588	-3.7321	-0.2679	3.8637	-1.0353
286.0	4.9916	-0.9613	0.2756	-3.4874	-0.2867	3.6280	-1.0403
287.0	5.0091	-0.9563	0.2924	-3.2709	-0.3057	3.4203	-1.0457
288.0	$8\pi/5$	-0.9511	0.3090	-3.0777	-0.3249	3.2361	-1.0515
289.0	5.0440	-0.9455	0.3256	-2.9042	-0.3443	3.0716	-1.0576
290.0	5.0615	-0.9397	0.3420	-2.7475	-0.3640	2.9238	-1.0642
291.0	5.0789	-0.9336	0.3584	-2.6051	-0.3839	2.7904	-1.0711
292.0	5.0964	-0.9272	0.3746	-2.4751	-0.4040	2.6695	-1.0785
293.0	5.1138	-0.9205	0.3907	-2.3559	-0.4245	2.5593	-1.0864
294.0	5.1313	-0.9135	0.4067	-2.2460	-0.4452	2.4586	-1.0946
295.0	5.1487	-0.9063	0.4226	-2.1445	-0.4663	2.3662	-1.1034
296.0	5.1662	-0.8988	0.4384	-2.0503	-0.4877	2.2812	-1.1126
297.0	5.1836	-0.8910	0.4540	-1.9626	-0.5095	2.2027	-1.1223
298.0	5.2011	-0.8829	0.4695	-1.8807	-0.5317	2.1301	-1.1326
299.0	5.2185	-0.8746	0.4848	-1.8040	-0.5543	2.0627	-1.1434
300.0	$5\pi/3$	-0.8660	0.5000	-1.7321	-0.5774	2.0000	-1.1547
301.0	5.2534	-0.8572	0.5150	-1.6643	-0.6009	1.9416	-1.1666
302.0	5.2709	-0.8480	0.5299	-1.6003	-0.6249	1.8871	-1.1792
303.0	5.2883	-0.8387	0.5446	-1.5399	-0.6494	1.8361	-1.1924
304.0	5.3058	-0.8290	0.5592	-1.4826	-0.6745	1.7883	-1.2062
305.0	5.3233	-0.8192	0.5736	-1.4281	-0.7002	1.7434	-1.2208
306.0	$17\pi/10$	-0.8090	0.5878	-1.3764	-0.7265	1.7013	-1.2361
307.0	5.3582	-0.7986	0.6018	-1.3270	-0.7536	1.6616	-1.2521
308.0	5.3756	-0.7880	0.6157	-1.2799	-0.7813	1.6243	-1.2690
309.0	5.3931	-0.7771	0.6293	-1.2349	-0.8098	1.5890	-1.2868
310.0	5.4105	-0.7660	0.6428	-1.1918	-0.8391	1.5557	-1.3054
311.0	5.4280	-0.7547	0.6561	-1.1504	-0.8693	1.5243	-1.3250
312.0	5.4454	-0.7431	0.6691	-1.1106	-0.9004	1.4945	-1.3456
313.0	5.4629	-0.7314	0.6820	-1.0724	-0.9325	1.4663	-1.3673
314.0	5.4803	-0.7193	0.6947	-1.0355	-0.9657	1.4396	-1.3902
315.0	$7\pi/4$	-0.7071	0.7071	-1.0000	-1.0000	1.4142	-1.4142

Appendix B Continued

TRIGONOMETRIC FUNCTIONS

Degrees	Radians	Sine	Cosine	Tangent	Cotangent	Secant	Cosecant
316.0	5.5152	-0.6947	0.7193	-0.9657	-1.0355	1.3902	-1.4396
317.0	5.5327	-0.6820	0.7314	-0.9325	-1.0724	1.3673	-1.4663
318.0	5.5501	-0.6691	0.7431	-0.9004	-1.1106	1.3456	-1.4945
319.0	5.5676	-0.6561	0.7547	-0.8693	-1.1504	1.3250	-1.5243
320.0	$16\pi/9$	-0.6428	0.7660	-0.8391	-1.1918	1.3054	-1.5557
321.0	5.6025	-0.6293	0.7771	-0.8098	-1.2349	1.2868	-1.5890
322.0	5.6200	-0.6157	0.7880	-0.7813	-1.2799	1.2690	-1.6243
323.0	5.6374	-0.6018	0.7986	-0.7536	-1.3270	1.2521	-1.6616
324.0	$9\pi/5$	-0.5878	0.8090	-0.7265	-1.3764	1.2361	-1.7013
325.0	5.6723	-0.5736	0.8192	-0.7002	-1.4281	1.2208	-1.7434
326.0	5.6898	-0.5592	0.8290	-0.6745	-1.4826	1.2062	-1.7883
327.0	5.7072	-0.5446	0.8387	-0.6494	-1.5399	1.1924	-1.8361
328.0	5.7247	-0.5299	0.8480	-0.6249	-1.6003	1.1792	-1.8871
329.0	5.7421	-0.5150	0.8572	-0.6009	-1.6643	1.1666	-1.9416
330.0	$11\pi/6$	-0.5000	0.8660	-0.5774	-1.7321	1.1547	-2.0000
331.0	5.7770	-0.4848	0.8746	-0.5543	-1.8040	1.1434	-2.0627
332.0	5.7945	-0.4695	0.8829	-0.5317	-1.8807	1.1326	-2.1301
333.0	5.8119	-0.4540	0.8910	-0.5095	-1.9626	1.1223	-2.2027
334.0	5.8294	-0.4384	0.8988	-0.4877	-2.0503	1.1126	-2.2812
335.0	5.8469	-0.4226	0.9063	-0.4663	-2.1445	1.10338	-2.3662
336.0	5.8643	-0.4067	0.9135	-0.4452	-2.2460	1.0946	-2.4586
337.0	5.8818	-0.3907	0.9205	-0.4245	-2.3559	1.0864	-2.5593
338.0	5.8992	-0.3746	0.9272	-0.4040	-2.4751	1.0785	-2.6695
339.0	5.9167	-0.3584	0.9336	-0.3839	-2.6051	1.0711	-2.7904
340.0	$17\pi/9$	-0.3420	0.9397	-0.3640	-2.7475	1.0642	-2.9238
341.0	5.9516	-0.3256	0.9455	-0.3443	-2.9042	1.0576	-3.0716
342.0	$19\pi/10$	-0.3090	0.9511	-0.3249	-3.0777	1.0515	-3.2361
343.0	5.9865	-0.2924	0.9563	-0.3057	-3.2709	1.0457	-3.4203
344.0	6.0039	-0.2756	0.9613	-0.2867	-3.4874	1.0403	-3.6280
345.0	6.0214	-0.2588	0.9659	-0.2679	-3.7321	1.0353	-3.8637
346.0	6.0388	-0.2419	0.9703	-0.2493	-4.0108	1.0306	-4.1336
347.0	6.0563	-0.2250	0.9744	-0.2309	-4.3315	1.0263	-4.4454
348.0	6.0737	-0.2079	0.9781	-0.2126	-4.7046	1.0223	-4.8097
349.0	6.0912	-0.1908	0.9816	-0.1944	-5.1446	1.0187	-5.2408
350.0	6.1087	-0.1736	0.9848	-0.1763	-5.6713	1.0154	-5.7588
351.0	6.1261	-0.1564	0.9877	-0.1584	-6.3138	1.0125	-6.3925
352.0	6.1436	-0.1392	0.9903	-0.1405	-7.1154	1.0098	-7.1853
353.0	6.1610	-0.1219	0.9925	-0.1228	-8.1443	1.0075	-8.2055
354.0	6.1785	-0.1045	0.9945	-0.1051	-9.5144	1.0055	-9.5668
355.0	6.1959	-0.0872	0.9962	-0.0875	-11.430	1.0038	-11.474
356.0	6.2134	-0.0698	0.9976	-0.0699	-14.301	1.0024	-14.336
357.0	6.2308	-0.0523	0.9986	-0.0524	-19.081	1.0014	-19.107
358.0	6.2483	-0.0349	0.9994	-0.0349	-28.636	1.0006	-28.654
359.0	6.2657	-0.0175	0.9998	-0.0175	-57.290	1.0002	-57.299
360.0	2π	0.0000	1.0000	0.0000	undefined	1.0000	undefined

ANSWERS TO ODD-NUMBERED EXERCISES

Set 2.1

1.

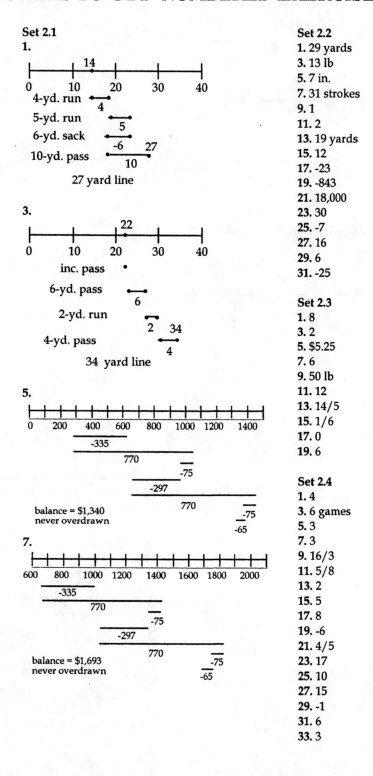

3.

5.

7.

Set 2.2

1. 29 yards
3. 13 lb
5. 7 in.
7. 31 strokes
9. 1
11. 2
13. 19 yards
15. 12
17. -23
19. -843
21. 18,000
23. 30
25. -7
27. 16
29. 6
31. -25

Set 2.3

1. 8
3. 2
5. $5.25
7. 6
9. 50 lb
11. 12
13. 14/5
15. 1/6
17. 0
19. 6

Set 2.4

1. 4
3. 6 games
5. 3
7. 3
9. 16/3
11. 5/8
13. 2
15. 5
17. 8
19. -6
21. 4/5
23. 17
25. 10
27. 15
29. -1
31. 6
33. 3

Set 2.5
1. $200 < d < 220$
3. $0 \leq f \leq 3$, f an integer
5. $3 \leq g \leq 9$
7. $0° < A < 180°$
9.

11.

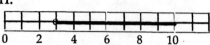

13.

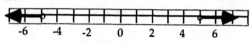

15.

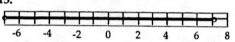

17.

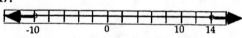

19. $x > 6$
21. $x < -26$
23. $x < -5$

Set 3.1
1. 120 lb
3. 160 lb
5. 250,000 lb

Set 3.2
1. A.) 15,000 lb/in.^{2}, B.) 0.5 in.^{2}, C.) 2,000 lb/in.^{2}, D.) 500 lb, E.)2.5 in.^{2}
3. 4 in.
5. $\sqrt{ab}$
7. $\dfrac{\pi s^2}{4}$
9. 10.1m

Set 3.3
1. a.) 3,000 lb/in.^{2} , b.) 12,000 lb/in.^{2}
3. $\sqrt[3]{\dfrac{2bh^2}{3\pi}}$
5. a.) M_A = 213,333 in.·lb, b.) M_B = 426,667 in.·lb
7. 3
9. ±7
11. 5

13. ±0.4
15. 0.4
17. 7
19. 2
21. 5
23. 121.5
25. $x^2 y^{-8}, x \neq 0 \quad y \neq 0$
27. $x^7, \quad x \neq 0$
29. $\dfrac{x^2}{y^3}, y \neq 0$

Set 3.4
1. 20.6 in.
3. a.) 503 ft.^{3}, b.) 30,159 lb.
5. V_w = 285 in.^{3}, W_w = 86 lb
7. 2 in.

Set 3.5
1. 2.46×10^{-5}
3. 9.41×10^8
5. -9.4221×10^4
7. 4×10^{-7}
9. 209,000
11. 524.3
13. -0.000345
15. 90.08
17. -2.06×10^4
19. -1.8×10^{-1}
21. 6×10^{-4}

Set 4.1
1. a.)155 yd., b.) 9.7 yd.
3. 0.400
5. a.) 196, b.) 203

Prob.	mean	med.	max.	min.	range
7.	29.3	35	40	4	36
9.	162	211.5	304	27	277
11.	15	12	34	1	33

Set 4.2
1.

	Batting Averages in Game:			
Player	1	2	3	4
Crane	0.500	0.600	1.000	0.333
Franco	0.167	0.400	0.600	0.000
Lamb	0.400	0.667	0.750	0.333
Carr	0.000	0.333	0.333	0.800
Owens	0.750	0.000	0.500	0.333
Clifford	0.200	0.500	1.000	0.500
Griffith	1.000	1.000	0.333	0.667
Mendez	0.400	0.750	0.750	0.750
Davis	0.000	0.800	0.667	0.667

Team Totals				
Game	Hits	Walks	At-bats	Batting Ave.
1	15	7	49	0.357
2	22	11	50	0.564
3	23	15	51	0.639
4	15	20	48	0.607

3.

Team	Pct. Win	Goals/Game
Pittsburgh	0.636	1.2
Montreal	0.568	1.1
Boston	0.559	1.1
Buffalo	0.345	1.3
Quebec	0.436	0.9
Hartford	0.415	0.8
Ottawa	0.190	0.4

5.

Player	3-Point %	2-Point %	Free Throw %	Total Points
Unruh	0.500	0.529	0.833	44
Rollins	1.000	0.400	0.500	29
Farris	0.333	0.444	1.000	13
Driscol	0.800	0.545	0.000	24
Lang	0.833	0.750	0.857	45
Keefe	0.250	0.444	0.444	23
Edwards	0.667	0.667	0.667	12
Graham	1.000	0.500	1.000	26
Worthy	0.000	0.000	0.000	18
Campbell	0.000	0.533	1.000	18

Set 5.1

1.

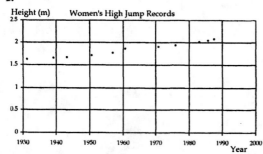

3. (-4,2), II
5. (1,4), I
7. (-2,0)
9. (2,-1), IV
11. (0,-3)
13.

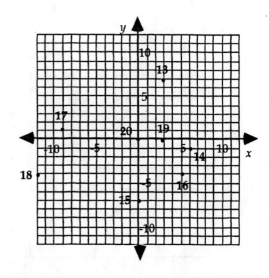

Set 5.2
1. 1st base, 72.8 ft.; 2nd base, 107.7 ft.; 3rd base, 71.3; home 17.1
3. 10
5. 58
7. 14.4

Set 5.3
1. A(4,0,-0.2), B(0.5,0,-0.7), C(3,1,-1.2), D(3,2,-0.6), E(0,1,-1.8), F(4.5,4,-1.1), G(1,3,-2.1), H(2,5,-0.9), I(0,4,-1.7)
3. 5.2
5. 36.8

Set 5.4
1. 30 in.
3. h_u = 35.7 in., M_u = 714 in·lb; h_c = 27.8 in. M_c = 556.5 in·lb
5.

Component	w	x·w	z·w
Wing	9,315	321,368	111,780
Body/Tail	17,808	633,965	165,614
Eng./Prop.	4,216	118,048	44,268
Fuel	20,183	696,314	242,196
Payload	26,828	912,152	187,796
Totals	69,043	2,681,846	751,654

(34.2,0,9.6)

Set 6.1
1. 36m
3. 0.56 sec.
5. 0.72 sec.
7. faster than 95.5 ft./sec
9. 18 ft.
11. 15 ft./min.
13. 6,000 in.·lb
15. 15 sec.

Set 6.2

1.

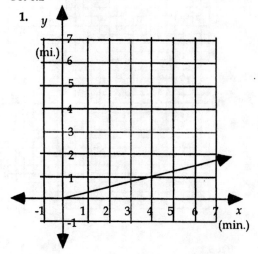

5.

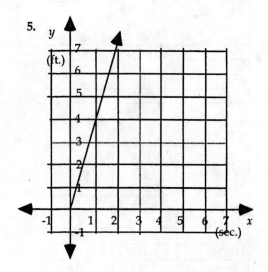

3.

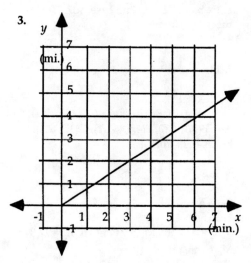

7.

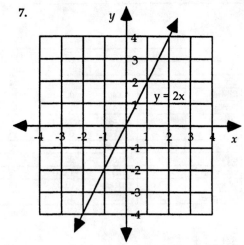

9.

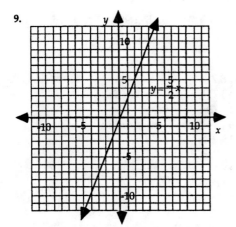

17.

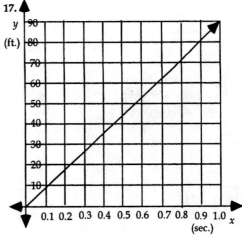

11.

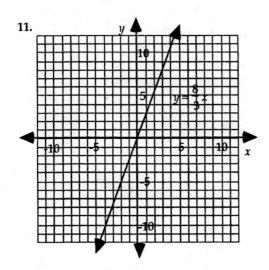

19.

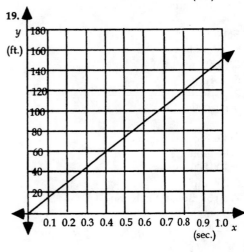

15.

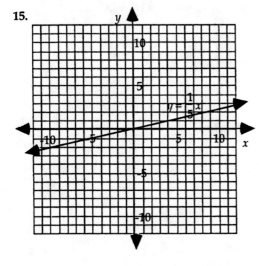

Set 6.3
1. $y = 140$
3. $x = 2.7$
5. $5/2$
7. $4/5$
9. $1/8$
11. 0
13. -1
15. -3
17. 0

Set 6.4
1. 3.5 sec.
3. 0.5 sec.
5. $y = 7$
7. $x = 12$
9. $y = 265$
11. $y = 1,050.5$
13. $x = \sqrt{5}$
15. $y = 0$
17. $y = 1$
19. $y = \sqrt{21} - 5$

21. -2/3
23. no slope
25. 0

Set 6.5
1. 7 yard line
3. 80 yards (other 20 yard line)
5. 2 yard line (never catches)
7. Debbie passes opponent 51m after receiving the baton.
9. (-6,-40)
11. no solution
13. (-33,-16)
15. (1/0.7,0)
17. $\left(\sqrt{2}, 6\right)$
19. no solution

Set 6.6
1. 9 yard line
3. (5,7)
5. (3,4)
7. (-1,1)
9. (12,27)

Set 6.7
1. 80 mph
3. 100,000 papers/hr.
5. 330 mph
7. a.) 30 mph, b.) tailwind
9. 33 mph
11. 2.4 days

Set 6.8
1. yes
3. -1/4
5. a.) $y_5 = -\dfrac{3}{2}x + 5$, b.) yes
7. 468 ft.

Set 7.1
1. 5 touchdowns, 3 field goals, 2 safeties
3. (0,5,-3)
5. (3,7,-1)
7. (2,-1,3)

Set 7.2
1. first 10.12 sec, second 10.48 sec., third 10.3 sec.
3. (0,5,-3)
5. (-1.2,3,0)
7. (2,-1,3)

Set 8.1
1. a.) 90i, b.) 90i + 90j, c.) 90j, d.) 0i + 0j, e.) 90j, f.) -90i, g.) -90j

3. a.) G & 2nd, b.) B & 9th, c.) A & 3rd, d.) F & 9th, e.) H & 4th, f.) D & 4th, g.) B & 2nd, h.) E & 6th

Set 8.2
1. a.) -29i + 211j, b.) -1i + 2j
3. a.) -26i - 24j, yes
5. a.) 44i - 39j, b.)59 ft./sec
7. 15i + 2j

Set 8.3
1. 44 ft./sec, 30 mph
3. 12.7 rpm
5. 0.068
7. a.)-θk, b.) θk, c.) -θj, d.)θj, e.)θi, f.)-θi

Set 8.4
1. F_2 = 0i - 40j - 30k
3. a.) 146.3i + 2.1j + 6.5k ft./sec., b.) 146.5 ft./sec
5. 560.4i + 15j +13k in./sec.
7. 60i - 48j
9. 0i + 0j + 28k
11. 6i + 24j - 12k
13. 0i + 3j + 0k

Set 9.1
1. a.) 1.04 sec., b.) 10.2 m/sec.
3.

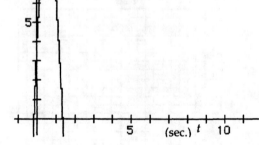

$$y = -16.1t^2 + 20t + 3$$

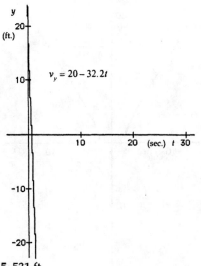

$v_y = 20 - 32.2t$

5. 531 ft.

7. 10 ft.

9. 46.8 ft./sec

11. a.) 34 ft., yes, b.) 38.8 ft., c.) 223.6 ft., d.) $y = -0.0031x^2 + 0.69x$, e.)

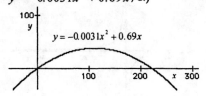

$y = -0.0031x^2 + 0.69x$

13. a.)

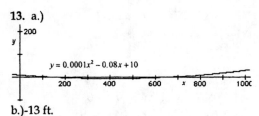

$y = 0.0001x^2 - 0.08x + 10$

b.)-13 ft.

Set 9.2

1. 2.5 sec.

3. 10 ft.

5. -3

7. $\{1,5\}$

9. $\{6,8\}$

11. $\{-7,-6\}$

13. $\{-4,-3\}$

15. $\{-6,7\}$

17. $\{-7,3\}$

19. $\{-9,5\}$

21. $\{-9,2\}$

23. $\{-2,3\}$

25. $\left\{\dfrac{2}{3},2\right\}$

27. $\{-1.5,-1\}$

Set 9.3

1. 52 yd.

3. 100 and 150 ft.

5. 5.7 in.

7. $\left\{-\dfrac{7}{2},6\right\}$

9. $\{-1,4\}$

11. $\left\{-\dfrac{1}{2},3\right\}$

13. $\{1,3\}$

15. $\left\{-3,\dfrac{1}{2}\right\}$

Set 9.4

1. a.) (0,4), b.) $y = -4$

3. 12 in.

5. 12 in.

Set 9.5

1. 3.8 in.

3. (23.8,97.1)

5.

x (ft.)	Diameter (ft.)
-60	0
-50	11.06
-40	14.91
-30	17.32
-20	18.86
-10	19.72
0	20
10	19.36
20	17.32
30	13.23
40	0

7.

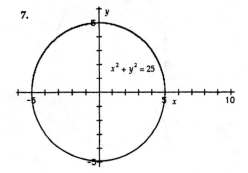

$x^2 + y^2 = 25$

9.

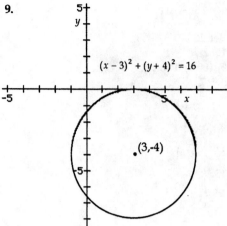

$(x - 3)^2 + (y + 4)^2 = 16$

(3,-4)

11.

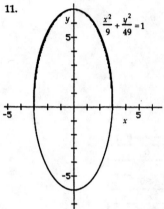

$\dfrac{x^2}{9} + \dfrac{y^2}{49} = 1$

13.

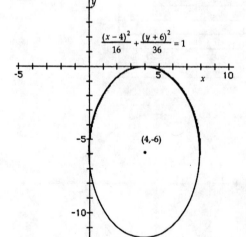

$\dfrac{(x - 4)^2}{16} + \dfrac{(y + 6)^2}{36} = 1$

(4,-6)

15.

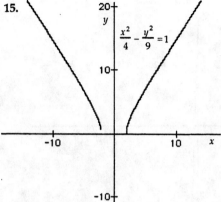

$\dfrac{x^2}{4} - \dfrac{y^2}{9} = 1$

17.

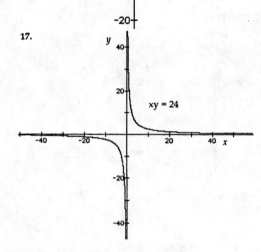

$xy = 24$

Set 10.1
1. $-x^4 + 2x^3 + 5x^2 - 7x - 6$
3. $-6y^2 + 4y - 2xy - 4x^2 - 9x$
5. $3x^{\frac{1}{2}} + 2x^{-\frac{1}{2}} + 8x^{-\frac{3}{2}}$
7. $5x\sqrt[3]{y} - x^3 - x^2 + 7$
9. $-4y^2 + 3xy - 3x^2 - 31$
11. $22x^5 - 7x^3 - 9x^2 + 6x - 4$

Set 10.2
1. $x^2 + 3x - 28$
3. $x^3 + x^2y - 7x - 7y$
5. $x^2 + 3x^{\frac{3}{2}} - 4x$

Set 10.3
1. $\dfrac{x - 7}{x + 6}$ $\qquad x \notin \{-6, -3\}$
3. $x + 1$ $\qquad x \notin \{0, 1\}$

5. $\dfrac{x^2 + x}{x^2 - x - 20}$ $x \notin \{-4,5,6\}$

Set 10.4

1. $\dfrac{x-2}{x^2 - 2x + 4}$ $x \neq 2$

3. $\dfrac{u^3 + u^2 v + uv^2 + v^3}{u^2 + uv + v^2}$ $u \neq v$

5. $x + 3$ $x \neq -3$

Set 10.5

1. $\dfrac{x^2 - 7x}{x + 4}$ $x \notin \{-4,1\}$

3. $\dfrac{xy - x}{xy - y + 5x - 5}$ $y \notin \{-5,1\}, x \neq 1$

5. $\dfrac{2x + 14}{x^2 - 8x - 15}$ $x \notin \{3,4,5\}$

7. $\dfrac{x^2 - 6x + 49}{x^2 - 6x - 7}$ $x \notin \{-1,7\}$

Set 11.1
1. 0.553
3. 0.612
5. 3
7. 2.93
9. 7.83
11. 9,140
13. 35,700
15. 637
17. 1,410,000

Set 11.2
1. 523 organisms
3. a.) 22 years, b.) 3.07×10^9 years

Set 12.1
1. a.) 2.7 yd., b.) 7.5 yd., c.) 7.5i - 2.7j
3. a.) 9.6 ft., b.) 12.6 ft., c.) 19.6 ft.
5. 45i + 9.6j
7. 0.04i + 0.005j
9. 12i - 7.2j
11. -274i - 122j
13. 99.9i + 3.5j
15. $x = vt\cos\theta$

17. $y = -\dfrac{gt^2}{2} + vt\sin\theta$

19. $y = -\dfrac{g}{2(v\sin\theta)^2} x^2 + \dfrac{\sin\theta}{\cos\theta} x$

21. $R = \dfrac{2v^2 \cos\theta \sin\theta}{g}$

Set 12.2
1. $22.5° < \theta < 31.8°$
3. 25.5°
5. 7.3 at 74.1°
7. 61.6 at -103.1°
9. 80.5 at 2.0 radians

Set 12.3
1. 0.8 bounces per second
3. a.) 0.15 lb., b.) 9.5 Hz, c.) 60 rad./sec., d.)

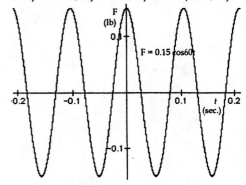

5. 1.4 in., 7.5 rad./sec.

INDEX